UNDERSTANDING CLIMATE CHANGE

Science, Policy, and Practice

Conversations about climate change are filled with challenges involving complex data, deeply held values, and political issues. *Understanding Climate Change* provides readers with a concise, accessible, and holistic picture of the climate change problem, including both the scientific and human dimensions.

Understanding Climate Change examines climate change as both a scientific and a public policy issue. Sarah L. Burch and Sara E. Harris explain the basics of the climate system, climate models and prediction, and human and biophysical impacts, as well as strategies for reducing greenhouse gas emissions, enhancing adaptability, and enabling climate change governance. The authors examine the connections between climate change and other pressing issues, such as human health, poverty, and other environmental problems, and they explore the ways that sustainable responses to climate change can simultaneously address those issues.

An effective and integrated introduction to an urgent and controversial issue, this book contains the tools needed for students, instructors, and decision-makers to become constructive participants in the human response to climate change.

SARAH L. BURCH is an assistant professor in the Department of Geography and Environmental Management at the University of Waterloo.

SARA E. HARRIS teaches in the Department of Earth, Ocean and Atmospheric Sciences at the University of British Columbia.

UNDERSTANDING CLIMATE CHANGE

Science, Policy, and Practice

SARAH L. BURCH AND SARA E. HARRIS

UNIVERSITY OF TORONTO PRESS
Toronto Buffalo London

Toronto Buffalo London
www.utppublishing.com
Printed in the U.S.A.

ISBN 978-1-4426-4652-0 (cloth)
ISBN 978-1-4426-1445-1 (paper)

Printed on acid-free paper with vegetable-based inks

Publication cataloguing information is available from Library and Archives Canada.

University of Toronto Press acknowledges the financial assistance to its publishing program of the Canada Council for the Arts and the Ontario Arts Council, an agency of the Government of Ontario.

University of Toronto Press acknowledges the financial support of the Government of Canada through the Canada Book Fund for its publishing activities.

Contents

Preface

After decades of negotiation, education, and contentious debate, climate change has pervaded the public consciousness. It has become a political weapon, a topic of dinner conversation, and a crucial entry point for discussing the multitude of ways in which many industrialized societies have become fundamentally unsustainable. As climate change gains popular visibility, however, it is clear that the most basic causes and consequences of the problem are frequently misunderstood, leaving crucial gaps in the search for solutions. Furthermore, because of the complex and constantly evolving nature of climate change, efforts to analyze it commonly shave off a thin slice and avoid interconnections: economic dimensions without consideration of the political implications, introductions to the scientific underpinnings without insight into the social dynamics that characterize responses. The result is that those interested in engaging in the discussion might lack the tools they need to characterize the problem adequately. Furthermore, critical opportunities to deliver on multiple priorities simultaneously may be missed if the true complexity of social and biophysical systems remains unexplored.

The goals of this book are threefold. First, we offer the fundamentals of climate change – both the biophysical and human dimensions – to equip both the novice and the experienced practitioner or scholar with a clear and concise depiction of this pressing challenge. Second, we interweave human responses and solutions with the scientific facets of climate change, to position mitigation and adaptation strategies within the context of planetary energy flows. Finally, we use the lens of transformative change, through which the abundance of climate change action plans, proposals, and campaigns can be examined, to reveal the potential of these actions to dramatically alter emissions pathways and vulnerability. The lens of transformative change also suggests that the challenge of climate change, in fact, presents a powerful opportunity for *improving* the social, economic, and environmental sustainability of our communities – a much more inspiring task than simply striving to avoid disastrous impacts.

This book can be used as an introduction for interested individuals, a tool for undergraduate and graduate students seeking background in either the science or socio-political dimensions of climate change, and a reference for experienced scholars and practitioners who want to be clear on the basics. We have designed the sections geared towards policy and social issues to withstand the rapid pace of change in this arena, while the chapters that explore the science of climate change provide timeless fundamentals.

We would like to thank the many colleagues, friends, and supporters who have made this book possible, including Phil Austin, Mark Burch, Dwayne Bryshun, Jordan Dawe, Beth Doxsee, Tom-Pierre Frappé-Sénéclauze, Emily Harris, Robert Hunter, William Koty, Nathan Lauster, Leah MacFadyen, Trevor Murdock, Barry Norris, Collin Oldham, Sabina Oldham, John Robinson, Alison Shaw, and Stephen Sheppard. We would also like to thank our collegial communities at the University of British Columbia and the University of Waterloo, who support the crossing of interdisciplinary boundaries and deep engagement with civil society and communities of practice.

Finally, we would like to thank Jennifer DiDomenico and the excellent team at the University of Toronto Press, as well the reviewers of our early drafts, for their patient and thorough support.

UNDERSTANDING CLIMATE CHANGE

Science, Policy, and Practice

CHAPTER ONE

Climate Change in the Public Sphere

1.1. Communicating about climate change
1.2. The state of the science
1.3. Responding to climate change: Mitigation and adaptation
1.4. A brief history of climate change policy
- 1.4.1. The United Nations Framework Convention on Climate Change and the Kyoto Protocol
- 1.4.2. The Intergovernmental Panel on Climate Change

1.5. The scale of the challenge: Accelerating action on climate change
1.6. Roadmap of the book

MAIN POINTS:

- All around us we see evidence of the human impact on climate, with serious implications for environmental, economic, and social sustainability.
- Communicating the need to address climate change requires an understanding of the key stakeholder groups at play: the public, government, industry, scientists, and civil society.
- Tackling the climate change challenge requires the creation of a compelling vision of a desirable future, not just recapturing a mythical past or "tinkering around the edges" of our current development path.
- International climate change policy has been dominated by the United Nations Framework Convention on Climate Change and the work of the Intergovernmental Panel on Climate Change.
- Humans can respond to climate change either through mitigation (dealing with the causes) or adaptation (dealing with the effects).
- More data or science alone will not change individual behavior, but they are crucial for evidence-based decision-making.

From around the globe come the reports: rising sea levels and eroding coral islands in the Maldives, increasing hurricane activity affecting the US eastern seaboard and Gulf coast, drought in central Africa, declining amphibian populations in the Amazon. Accompanying these alarming reports, however, are stories of innovation, including promising developments in the creation of jet fuel from algae, zero-waste and carbon-neutral experimental communities in the United Arab Emirates, and grassroots community initiatives such as the Transition Town movement. The common link that connects these phenomena is the growing evidence that humans are having a dramatic impact on Earth's climate. Making sense of it all is a challenging task. How serious are the near-term effects of climate change? What will the world look like in 2100? Who should pay to solve the problem? Why do interminable international negotiations rarely seem to pay off with results?

The dominant rhetoric of the climate change conversation has emerged out of the modern environmental movement, which has been growing and reinventing itself since the early 1960s.[1] As evidence has mounted of accelerating deforestation, the widespread effects of toxic pesticides and industrial emissions on human health and ecosystems, a growing list of endangered species, and exploding human populations, the conversation at the core of the environmental movement largely (and understandably) has become focused on the damaging effect of human activity.[2]

The theme of limits – of modern development as a scourge or virus that has run rampant over the planet – has pervaded global consciousness and is fundamentally shaping our response to environmental challenges.[3] Furthermore, we have become fixated on defining, calculating, and predicting the exact extent to which humans have exceeded these limits. There is no denying, as we shall see in this book, that the challenges are indeed monumental, but these inherently negative messages are having two major consequences.[4] First, a common response to a message of unavoidable and impending apocalypse is often one of apathy, disempowerment, and shame. How did we let ourselves get to this point? Is human civilization simply an experiment in mass greed, recklessness, and violence? If so, and if our ultimate demise is virtually inevitable, what good is merely "tinkering around the edges" to try to fix it?

The second consequence, intimately related to the first, is that this apocalyptic framing suggests that human behavior needs to be constrained and managed so as to do slightly less harm. But what if we want more than just to recapture the "healthier" planet of two hundred years ago, but instead to *improve* well-being, health, equity, community, and a host of other factors? The framing that currently dominates the climate change discourse, however, reduces the likelihood of a focus on a creative, positive, nuanced vision of the future – one that is rooted in a deep scientific understanding of Earth systems but that also captures (or at least begins a conversation about) core human values such as equity, compassion, innovation, and connection.[5]

In this book, we offer a deeper understanding of the state of climate change science, and develop a profoundly human picture of the opportunity that now exists to create a healthier, more equitable, more resilient future. This is an entirely different task than recapturing some mythical past when humans lived in harmony with the Earth (and each other), or simply minimizing our destructive impact on the planet. In this first chapter, we introduce a few of the core concepts needed to delve into the science and policy of climate change. We explore the scientific community's broad findings of humanity's influence on the climate system, the policy tools available to respond to this challenge, and the core of the climate change debate in the scientific, political, and public communities.

1.1. Communicating about climate change

Since the late twentieth century, the issue of climate change has been taken up and debated by scientists, politicians, pundits, activists, and concerned citizens. The search for solid ground amid uncertainty and efforts to design effective and equitable policy responses to climate change have come to dominate the environmental agenda, pushing issues of water scarcity, deforestation, pollution, and species preservation deeper into the background of public consciousness.

All around us is evidence of climate change. It emerged first as a scientific issue, and scientists remain its most vocal messengers. Indeed, the debate is permeated with scientific terminology, which, some argue, has caused a disconnect between our understanding of the problem and the reality of the challenge we face: the causes and effects of, and solutions to, climate change are as much about values as about science. Scientists, furthermore, often lack the communication skills to engage meaningfully with public and policy-maker audiences, which inhibits the effectiveness of evidence-based decision-making. Casting climate change as a values-based issue, however, brings its own challenges. A moral framing of climate change brings to the fore cultural differences (such as varying perceptions of human rights), deeply entrenched political pathways, and religious beliefs. At the heart of the climate change debate are fundamental questions:

- Is the current mode of development in the West (and elsewhere) sustainable?
- What right does any one group have to affect the well-being of current and future generations?
- Who should pay to solve a problem – those who created it in the first place or those who are most likely to contribute to it in the future?
- How certain must we be about the science before we take precautionary action?

Note that only the last (and elements of the first) of these questions is scientific; the others are ethical, political, economic, and even spiritual. It is for this reason that the climate change debate is so fierce. The debate, indeed, defies simplification. One cannot assume that those who argue for action on climate change understand the science and are demanding the implementation of policy on a purely rational or altruistic basis. Similarly, one cannot assume that those who question taking action are unfamiliar with the science, influenced by vested interests, or lack concern for current and future generations. There are legitimate questions about the ability of models to predict future climatic shifts, and the most effective suite of policy responses has certainly yet to be found. Nevertheless, it is important to recognize that very high levels of certainty exist among scientists in the field about the fundamental conclusion that humans are affecting the climate.[6]

The challenge of climate change draws together stakeholders with both consistent and inconsistent values, goals, and priorities. None of these stakeholder groups is homogeneous, and thus they are difficult to characterize. For instance, while "the public" is often considered a major stakeholder, this group encompasses not only those who might have suffered from the dramatic effects of climate change (such as sea level rise or extreme weather events) but also those whose lifestyles have benefited enormously from the consumption of fossil fuels, as well as those who are informed and active in the climate change debate, those with little access to accurate scientific information, and those with no inclination to participate in decision-making processes, including elections.

Similarly, industry stakeholders include those with vested interests in perpetuating an economy rooted in fossil-fuel-based energy and those who stand to benefit from a transition to renewable technologies. Although some companies feel it is critical to show leadership on climate change as part of socially responsible business practice, others disagree with proposals to reduce greenhouse gas emissions or feel it is not their responsibility to act.

Governments, too, are playing a key role in the climate change debate, especially because most recent climate change policy has developed at the international level. This, however, generally excludes individuals and even non-state actors (such as civil society organizations and the private sector), leaving climate change decisions to be made by political elites. Governments also clearly have an important role to play in implementing legislated greenhouse-gas-reduction policies and sustainable land use plans and in supporting research into renewable energy. Even so, governments are subject to constantly changing political pressures, broader geopolitical conflicts, and term limits that influence the extent to which they are willing or able to plan for the long term.

The scientific community obviously plays an important role in our perception of, and response to, climate change, but that role is shifting. Scientists traditionally have been viewed as "objective" or value-neutral parties who simply communicate discoveries to policy-makers and the public, but it has now become clear that science itself is a value-laden practice. Decisions about which scientific research projects deserve funding are often shaped, in part, by the political context. Some governments funnel spectacular quantities of cash into the development of clean energy technologies or infrastructure design projects, while perceiving little use for studies of behavioral change, politics, and policy design. In addition, scientists often forge professional collaborations on the basis of personal relationships and perceived demand for particular work. This is not to claim that value-based aspects of science necessarily create bias in outputs or conclusions; rather, it highlights that the objective and subjective worlds are deeply intertwined and serve to produce knowledge that must be placed in the human context from which it emerges. Navigating these waters carefully is a crucial task, and it is certainly not an attractive option to leave the scientific community on the sidelines of decision-making. If we are to face the challenge at hand, leading-edge science on the causes and consequences of climate change must be fed into the decision-making process.

Another key stakeholder or actor that has emerged in the climate change debate is the media. If climate change was once the sole domain of scientists and science journalists, it is now explored by political, economic, security, and even fashion writers, with dramatic implications for the framing of the issue. Most obviously, the sensational elements and emotional stories associated with climate change are the first to be picked up and publicized by the media: polar bears and their sensitivity to diminishing sea ice have become the media's symbol of climate change, as have shocking storms or raging wildfires. Social scientists Max and Jules Boykoff have explored the ways in which the media influence the framing of the climate change debate and the messages the public consumes.[7] A core tenet of responsible reporting is balance; typically, this means giving equal "air time" to both sides of an issue. In the case of global warming, however, this has created an unintentional form of bias in which the opinions of a small group of skeptics are given weight equal to those of the vast majority of scientists, who have found climate change to be a real and valid concern. This bias dramatically misrepresents the state of scientific consensus on climate change.

Acknowledging that climate change is not simply a scientific issue but one that passes through the filter of human psychology raises the issue of risk perception, the study of which has burgeoned with escalating public concern about issues such as nuclear power, pollution, and natural disasters. What do we think are the risks from climate change? How important is it to respond to them? Should we respond to them before they occur or after? These questions reveal the multitude of ways in which such risks are perceived, and help to explain varying levels of action.

Humans' perception of the magnitude and likelihood that a risk will affect their lives is the product of a constellation of factors, many of which are relatively unrelated to the actual objective nature of the risk itself. Many hazards exist: air pollution, noise, and radiation, for instance, and the effects of climate change, such as extreme weather events, drought, and rising sea levels, are hazards as well. In their day-to-day activities, individuals might not come in contact with these hazards – they might live in a wealthy city, for example, away from a coastline or floodplain – but others might make a living from farming, or live in vulnerable low-lying areas prone to floods, or far from the disaster-management or public health resources that some cities have at their disposal. Many demographic characteristics also feed into this matrix, helping to determine both the activities that put us in the path of the hazard and the perception we have of it. Studies show that socially marginalized populations often perceive risks to be greater than do those who are relatively empowered.[8] Furthermore, perception of risk is directly influenced by individuals' worldviews and values, and the level of knowledge they possess. Research shows that the more scientifically literate people are, the more *polarized* they are in their opinions about climate change.[9] In other words, individuals' beliefs about risks are formed by what is appropriate in their community – their family, friends, the political groups they associate with – rather than strictly based on facts, figures, and scientific information.

The complications do not end there, of course. Even when individuals perceive a risk to be severe, they might not act to protect themselves or prevent the risk from occurring. Climate change is a testament to this paradox. Human behavior is a complicated phenomenon, one that we desperately need to understand better. Although the focus traditionally has been on educational campaigns to improve people's understanding of climate science or the need to respond to climate change, the scholarly community generally believes that this approach is dramatically insufficient to yield sustained shifts in behavior. Underlying values play a critical role in determining the effectiveness of information campaigns, as do beliefs about whether or not humans are at risk.[10] In the end, human beings are rarely simply rational creatures. Our emotions play a central role in determining which behaviors we choose,[11] and the most effective campaigns to persuade people to change their behavior recognize this. Furthermore, the context in which individuals go about their daily lives might present significant external barriers to behavioral change. For instance, they might value environmental sustainability and perceive significant risks associated with climate change, but be unable to make the shift from driving cars to using public transit if it is unavailable or inconvenient.

It is important to keep these complex drivers of human behavior in mind as we learn more about the fundamentals of climate science. In the interest of injecting a little sanity into the climate change debate, this book highlights what scientists are certain about, as well as areas where the science is still rapidly evolving. These distinctions are crucial to making informed decisions about response options, and form the context within which we can start to have an informed conversation about what sort of future we desire.

1.2. The state of the science

Earth is getting warmer, the past century of heating (about 0.8°C) is mostly due to increases in atmospheric greenhouse gases, and these increases are due to human activity. These statements have high scientific certainty, and are the core of the conclusion that human activities influence Earth's climate. The evidence for rising global average temperature comes from many thousands of measurements, analyzed in different ways by different groups of scientists, all of whom have found very similar temperature patterns over time. There are no credible challenges to this evidence.

How do we know that greenhouse gases are the primary culprits? Climate has three basic controls: the Sun's incoming energy, reflection of the Sun's energy back into outer space, and the atmospheric greenhouse effect. Neither a change in incoming solar radiation nor a change in reflection can explain the recent warming, as we discuss in Chapters 3 and 4. An increase in the greenhouse effect is the only major control going in the same direction as the temperature observations. In addition, the lower atmosphere has warmed, while the upper atmosphere – the stratosphere – has cooled a little, which is consistent with more greenhouse gases keeping more energy in the lower atmosphere for longer. If the Sun were responsible, we would expect warming throughout the atmosphere, which is not what we observe.

How do we know humans are responsible? From historical records of fossil fuel extraction, we know how much fossil carbon we have burned and how much of that carbon has accumulated in the atmosphere. We have measured the chemistry of atmospheric carbon dioxide and found that the changes in chemistry match what would be expected from releasing carbon from fossil fuels and in the amounts that have been released. No known natural releases of carbon to the atmosphere could have done it. Climate models that include only natural variability – for example, changes in solar radiation and the discharges from volcanoes – cannot produce the observed temperature increase of the past century. Only by including human influences can the models simulate what has actually happened. Humanity's fingerprints on Earth's climate are clear.

What pieces do we know well? We have excellent information about the physical properties of greenhouse gases, how they absorb and re-emit energy, and how long they persist in the atmosphere. We have a good grasp of some of the climate responses that occur in a warming world, such as the behavior of water vapor (which enhances the heating) and how, as Earth warms, it emits more energy into outer space (which helps counteract the heating).

We know less about some other responses and drivers. How do clouds respond to warming? How much cooling do reflective droplets and particles in the atmosphere provide, offsetting some of the greenhouse heating? We also need better information about what to expect in specific places, such as coastlines and agricultural areas. Which crops will be well suited to the regional climates of the future? How often will flooding events occur and how severe will they be? Scientists are working to get a handle on these aspects of regional climate change.

Even with some imperfect pieces, however, the scientific questions around present-day climate change, on a human time scale, are not about whether climate is changing or even whether human actions influence climate, but about how fast and how much climate will change. Scientists will continue to test current thinking and fill in the gaps in the climate picture. Resolving the remaining questions will clarify the details, but it is unlikely to overthrow our current understanding of how Earth's climate works.

1.3. Responding to climate change: Mitigation and adaptation

Human systems and Earth systems are inextricably linked. The ways we choose to develop our cities, grow our food, and live our lives significantly influence both the quantity of emissions we put into the atmosphere and the degree of climate change we cause. Similarly, the places we choose to settle, and the way we build those settlements, directly influence our degree of vulnerability to the impacts of climate change. These connections suggest that there are two broad, and deeply interwoven, categories of responses to climate change: we can change the impact we have on climate, and/or we can minimize the impact that climate has on us.

Mitigation means getting at the "roots" of climate change. The goal is to prevent climate change before it starts or, at least, to reduce its effects once it has started. Mitigation has been the most common policy response to climate change since evidence of human interference with the planet's delicate climatic balance began to emerge. So, the most commonly held definition of mitigation refers to efforts to reduce greenhouse gas emissions or enhance the storage of carbon (the main culprit) in "carbon sinks" such as forests, oceans, and soil.

Adaptation involves strategies such as building higher dikes to keep out rising seas, growing crop varieties that thrive under warmer, or wetter, or drier conditions, and protecting vulnerable populations during extreme weather. Adaptation thus means tackling the consequences, rather than the causes, of climate change. As we shall see, adaptation is a challenging proposition when, as is common, those who will suffer the most severe impacts of climate change are likely not those who are responsible for causing it.

1.4. A brief history of climate change policy

Climate change is a complicated phenomenon to attempt to manage. It pervades and affects the very foundations of our way of life, requiring navigation through a maze of economic, scientific, societal, and governance concerns. Widely diverging interests collide in the international arena, and the risks associated with climate change are weighed against security threats, economic fluctuations, human rights debates, and geopolitical posturing. As such, the path towards legally binding global climate change legislation has been a bumpy one plagued initially by complex science and policy design, and later by significant controversy.

Over the past twenty years, the international community has toyed with mandatory greenhouse gas emissions levels for developed countries, voluntary limits for developing countries, mechanisms by which funds can be transferred from rich to poor to stimulate the growth of green technology, and funds to sponsor the protection of communities against climate change impacts. These efforts have met with mixed success, as we shall see throughout this book.

The core justification for responding to climate change at the global level is simple. Greenhouse gases released anywhere on the planet act essentially the same way in the climate system: they distribute evenly and affect the global climate, not just the region in which they were emitted. This means that a metric tonne of greenhouse gas emitted in Canada has the same impact as a tonne emitted in China or Cameroon. Furthermore, the global economy is now so tightly interwoven that the repercussions of exchanging cheap, dirty technologies for more expensive but clean ones can reverberate in ways that are difficult to predict. Fossil fuels are currently cheap for two reasons. First, billions of dollars are poured into subsidies around the world to ensure an inexpensive supply of fuel[12] and to support local industries, and, second, our current transportation, heating, and electricity infrastructure has been created with fossil fuels in mind. We are now faced with inertia: it is currently less costly (in financial terms) to stick with what we have than to overhaul our systems in favor of renewable energy. We deal with this in greater detail in Chapter 6.

A related implication of the global nature of climate change is that the world's distribution of wealth is such that a small fraction of the population is responsible for producing most of the emissions that are driving global climate change. In other words, the actions of a wealthy and consumptive few are creating a potentially unstoppable chain of events that could transform the livelihoods of billions of people the world over.

Finally, international climate change policy might lead to more efficient and coordinated outcomes than efforts that take place in isolation from one another at the local level. For instance, it might be cheaper to replace extremely dirty or inefficient coal-fired power plants in developing countries, rather than to pay much greater sums to yield small efficiency improvements in already relatively clean technologies.

From the negotiation of the Kyoto Protocol – the first major international agreement aimed at managing greenhouse gas emissions and responding to climate change – to protests associated with recent climate change negotiations, it is clear that creating effective international climate change policy is a far from simple task. The complexity of the global geopolitical landscape translates into varying goals and pressures at the negotiating table, while the cost of emissions reductions runs counter to political priorities at home. A lack of understanding of what a low-emissions world might look like exacerbates these issues.

Added to these challenges are barriers to the implementation of policy once it has been agreed upon. Without a ruling international government, only political pressure and economic measures can be brought to bear on those who fail to meet their promised emissions-reduction targets. This creates a strong incentive for "free riding" – that is, reaping the reward of actions taken by others. For example, if you sneak into a music concert and enjoy the show without paying, you are free riding; if the other people at the concert also had not paid, the musicians would not have performed, but since the musicians' costs were covered by tickets fairly paid for by others, you are receiving a benefit for which you did not pay. Other examples of free riding are enjoying smooth roads and clean water without paying taxes or, more to the present point, receiving the "benefit" of a stable climate without reducing one's greenhouse gas emissions while others have taken steps to reduce their own emissions.

In the sections that follow, we introduce the central elements of international climate change policy: the Intergovernmental Panel on Climate Change, and the United Nations Framework Convention on Climate Change. We address these, along with subnational and local efforts to respond to climate change, in greater detail in Chapters 6, 9, 10, and 11.

1.4.1. The United Nations Framework Convention on Climate Change and the Kyoto Protocol

In 1992, at the Conference on Environment and Development in Rio de Janeiro, Brazil, UN members produced the United Nations Framework Convention on Climate Change (UNFCCC), which forms the backbone of global climate change policy. The Convention was created to manage emissions of greenhouse gases and the resulting climate change, and although the UNFCCC did not contain binding emissions-reduction targets, these were adopted in a subsequent and related treaty, the Kyoto Protocol. The UNFCCC currently has 195 parties, not all of which have both signed and ratified the Kyoto Protocol.[13] One of the most important functions of the UNFCCC is to hold periodic meetings of parties to the Convention. These meetings, called Conferences of the Parties (COPs), occur approximately every twelve months – it was during COP3, for instance, in 1997, that the Kyoto Protocol, the most significant piece of international climate change policy to grow out of the UNFCCC, was created.

The issue on the table at COP3 in Kyoto, Japan, was the creation of emissions-reduction targets for all developed countries that would ratify the Protocol and to establish mechanisms through which reductions could be stimulated in developing countries. The ultimate goal was to stabilize greenhouse gas emissions at a level that would prevent dangerous levels of climate change. The targets were to be met between 2008 and 2012, with a successor to the Kyoto Protocol to be negotiated during that period. This has been a contentious issue, however, as many scientists argue that the agreed reductions were too modest to mitigate human influence on the global climate effectively. Furthermore, debate has been sparked by the role of developing countries, most notably India and China, in future emissions reductions. Of the list of countries that have not ratified the Protocol, the most notable, even though it is a signatory of the UNFCCC, is the United States. Moreover, in 1990 – the base year upon which reduction targets in the Kyoto Protocol were set – the United States was responsible for approximately 30 percent of global greenhouse gas emissions. Canada ratified the Protocol, but it appears to be one of the least successful in meeting its Kyoto obligations, its emissions having increased by more than 26 percent since 1990, rather than diminishing by the 6 percent to which it agreed. Indeed, in December 2011, Canada's lack of progress was made manifest when the federal environment minister publicly announced that Canada would formally withdraw from the Kyoto Protocol.

More recent COPs have been held in Copenhagen, Denmark; Cancun, Mexico; Durban, South Africa; and Doha, Qatar. The key issue at stake in the Copenhagen COP, held in December 2009, was the negotiation of the agreement that was to take effect following the end of the Kyoto Protocol commitment in 2012. After two weeks of intense negotiations, however, the parties failed to deliver a successor to the Kyoto Protocol, producing instead a brief, non-binding political declaration, although the richer nations did agree to make the first formal financial commitment to help poorer nations adapt to climate change.[14]

Following the perceived failure at Copenhagen, optimism surrounding international climate change negotiations experienced a modest recovery, mainly due to success in reducing emissions from deforestation and forest degradation (REDD) and the development of an adaptation framework. Even so, the emissions-reduction commitments agreed to would have gone just *60 percent of the way to a 50-50 chance* of reaching the goal of limiting warming to 2°C, and by 2013 the future of the Kyoto Protocol was left undecided.[15]

The points of contention at Copenhagen, and the very modest successes at the COPs that followed, reflect the broader debate swirling around global climate change policy. In particular, many developed countries are concerned that greenhouse gas emissions cannot be managed effectively without the binding participation of key developing countries whose emissions threaten to dwarf those of the West in the not-so-distant future. Other issues include funds to support climate change adaptation in developing countries, whether emissions from shipping by sea and international air travel should be included, and how to encourage the parties to agree to deep and binding reduction targets in a future protocol. Indeed, the mixed success of international negotiations has led to the growing popularity of bilateral (two-party) or "club" (a handful of parties) negotiations. For examples, it appears that the United States may pursue climate policy through the Major Economies Forum, a club of seventeen members that together emit more than 85 percent of global greenhouse gas emissions.

1.4.2. The Intergovernmental Panel on Climate Change

A core element of international climate change policy-making is the advice provided by the hundreds of scientists who constitute the Intergovernmental Panel on Climate Change (IPCC). Created in 1989 by the United Nations Environment Programme and the World Meteorological Organization, the IPCC gathers together the world's leading climate scientists to produce reports based on the latest scientific research. In periodic assessment reports, the IPCC reviews progress on climate change science – including work on both adaptation and mitigation – and synthesizes the material for use during policy negotiations. The importance of the IPCC's work was acknowledged in 2007 when, in tandem with US politician and climate activist Al Gore, it was awarded the Nobel Peace Prize. The award brought unprecedented attention to the work of the IPCC and pushed the climate change issue to new heights of public awareness.

To produce its reports, the IPCC divides its efforts into three working groups, each of which addresses one major component of climate change. Working Group I explores the science of climate change, and assesses our understanding of the drivers of climate change, projected changes in the climate, and observed changes around the world. Working Group II assesses studies on adaptation and impacts, including those on human health, settlements, and ecosystems, among many other areas. Working Group III explores the critical question of mitigation – strategies to prevent climate change from occurring or becoming more severe. Each working group draws upon the expertise of a wide range of scientists from many disciplines who act as authors and/or reviewers of the assessment reports. Together, the working groups make up a plenary panel, which feeds findings to the UNFCCC.

The IPCC is the largest scientific collaboration of its kind in human history. Experts in the three working groups meet frequently to collect, assess, and synthesize the scientific findings produced in thousands of publications around the world – a uniquely rigorous scientific process. This work is voluntary, and the individuals who participate are selected through a nomination and review process. Care is taken to ensure that author teams are comprised of individuals from a variety of views, nationalities, and scientific backgrounds, although criticism nonetheless has emerged about gender imbalance and the relative lack of inclusion of scholars from developing countries. As the assessment reports come together, they are subjected to an intensive review and critique virtually unparalleled in the scientific community. Experts who are not IPCC authors play an integral role in evaluating the assessment reports, including supplying new studies to evaluate and critiquing conclusions, to which IPCC authors must respond.

Despite its extensive review and revision processes, the IPCC is not without its critics. Valid concerns have been raised, for example, about the future value of the IPCC. As well, because the IPCC seeks consensus, its findings are often portrayed as scientifically conservative, a characterization that has been reinforced by the emergence of new findings, such as the increasing rate of ice melt in the Arctic, shortly after the completion and publication of a major IPCC assessment. Critics also charge that the IPCC neglects outlying views. Although these views are not necessarily incorrect, the IPCC seeks to gather science that has been validated and replicated by multiple experts, and cannot reasonably include the views of all scientists in the field. Furthermore, the IPCC must be policy relevant, but not policy prescriptive: it can document the ways in which human activities are affecting the climate, and propose response options, but it cannot argue for a particular policy solution. This is becoming an increasingly challenging line to walk, however, as pressure builds to demonstrate which response strategies are working, to grapple with the scale of the challenge, and to accelerate the transition towards sustainability. Indeed, since the IPCC grew out of the United Nations, it cannot directly criticize the climate change policies (or lack thereof) of the sovereign nation-states that are members of the United Nations. But this has led to criticism that the IPCC is abdicating its essential role as a scientific body that ought to be driving action on climate change. These and other criticisms have stimulated healthy debate, and could shift the role that the IPCC plays in the future.

1.5. The scale of the challenge: Accelerating action on climate change

The question of an effective response to the very real problem of human-induced climate change is a complicated one, touching on our scientific understanding of the problem, inertia built into technologies, and deeply rooted values. Closely interwoven human and natural systems make it challenging to determine how best to respond. Can we design cities that are low carbon, resilient in the face of climate change, and foster healthy, vibrant communities? Can we devise strategies that unleash the potential of innovative social enterprises, channel the ingenuity of the private sector, and bring policies at all levels of government in line with one another? Can we resist the urge to hive off one particular part of the problem and consider it in isolation? In attempting to create a desirable future, can we have a conversation that is simultaneously about social justice, ecological integrity, and well-being?

As daunting as this task may seem, we already possess many of the tools we need. Throughout this book, we will learn about real strategies that have been put into practice around the world that can help to shift communities onto more sustainable development paths. With core climate science concepts in hand, we will see how these response strategies relate directly to the causes and consequences of climate change, and evaluate both their feasibility *and* desirability. Climate change presents an opportunity to transform the way society functions. This book attempts to pull together the pieces of the puzzle and to equip you with the tools you will need to engage meaningfully in this debate and, ultimately, to accelerate our progress towards sustainability.

1.6. Roadmap of the book

In this book, we provide the scientific basics of climate change and possible response options – with respect to both mitigation and adaptation. We avoid an exclusive focus on the science of climate change, since this approach does not allow for a nuanced understanding of the social implications. Similarly, exclusively political or economic analyses of climate change frequently neglect the underlying science in terms of thresholds, feedbacks, and the potential for abrupt change. Our approach, rather, is to see the problem through the unique lens of "transformative change": are current climate change response strategies putting us on the path towards a fundamental transformation of emissions trajectories and vulnerability? We aim to explore visions of the future that are positive and ambitious, and that respond to what we know about the way humans are influencing the climate.

We begin by exploring the key scientific concepts that are crucial to understanding climate change. In Chapter 2, we look at systems, including stocks, flows, and feedbacks. In Chapters 3, 4, and 5, we explore the influence of energy from the Sun, reflectivity, and the greenhouse effect on the planet's weather and climate, looking not just at the causes of climate change, but also potential mitigation solutions.

In Chapter 6, we focus on a core driver of climate change: the global energy system. We also explore some of the most powerful currently available mitigation strategies, innovative actions that are being taken around the world, and leading-edge ideas on tackling the roots of climate change. In Chapters 7 and 8, using climate models and scenarios, we examine the challenges inherent in trying to predict the future, and offer a glimpse at what climate change might look like in the decades ahead.

In the last third of the book, we explore the human and natural consequences of climate change. In Chapter 9, we examine the impacts of climate change on natural systems, and potential strategies (or those that natural systems might follow spontaneously) to respond to these impacts. We also consider ways in which adaptation might imply synergies or trade-offs with mitigation strategies discussed in earlier chapters. In Chapter 10, we focus on the implications of climate change for human systems and settlements, with a particular focus on development, social justice, and equity. We explore specific adaptation strategies, the heated debate surrounding the costs of adaptation, and the linkages between adaptation and mitigation.

Given what we now know about the science behind climate change and the strategies at our disposal for responding to it, in the final chapter we look forward to the frontier of innovative action on climate change. We finish with a glimpse into the future of climate change policy and a look at the most promising opportunities for both managing the effects of climate change and ensuring a rapid transition to a fundamentally low-carbon development path.

CHAPTER TWO

Basic System Dynamics

2.1. What is a system?

- 2.1.1. System parts and interactions
- 2.1.2. Stocks and flows
- 2.1.3. Feedbacks
- 2.1.4. Lags
- 2.1.5. Function or purpose

2.2. Earth's climate system: The parts and interconnections

- 2.2.1. Atmosphere, hydrosphere, biosphere, geosphere, and anthroposphere
  - *2.2.1.1. The atmosphere*
  - *2.2.1.2. The hydrosphere*
  - *2.2.1.3. The biosphere*
  - *2.2.1.4. The geosphere*
  - *2.2.1.5. The anthroposphere*
- 2.2.2. The ins and outs of Earth's energy budget
  - *2.2.2.1. Does what comes in go out?*
  - *2.2.2.2. Climate sensitivity: How much bang for your buck?*

2.3. Integrating systems, science, and policy

MAIN POINTS:

- Earth's climate results from a complex interconnected system of stocks, flows, feedbacks, and delays.
- Humans do not have particularly good intuition about how systems function; systems thinking takes deliberate practice.
- Processes in the atmosphere, hydrosphere, biosphere, geosphere, and anthroposphere all influence energy flows in Earth's climate system.
- Imbalances in energy flows drive climate change.

Our planetary system is a multidimensional web of interconnections. All the pieces of the web have a connection – some direct, some obscure – to Earth's climate. The land and water on Earth's surface, physical and chemical interactions in the atmosphere, the photosynthesis and respiration of plants, decision-making and actions taken by human communities, and even the ideas of individuals have influenced and will continue to influence Earth's climate. It is a multiway street. The climate itself also influences all the parts of the web, even the decisions of humans. In this chapter, we provide an introduction to how such systems work and how they can change over time. One of the large unknowns in forecasting future climate is future human actions. Thinking about climate in a "systems" context can give us insights into the effects of particular actions and the time it takes for Earth's climate system to respond. These two types of information are useful for mitigation and adaptation planning.

2.1. What is a system?

Imagine a bridge over a river, a coal-fired power plant that produces electricity, and a penguin colony. Now imagine the behavior of the bridge, power plant, and penguin colony over time. Cars flow onto the bridge, cross it, then flow off; traffic jams develop and ease. In the power plant, coal burns, electricity flows to light bulbs, and carbon dioxide (CO_2) flows into the atmosphere; equipment fails and is fixed. The penguins live and breed, their population going up and down over time, depending on factors such as food supply, predators, and the weather.

These three examples, and an infinite number of others – human bodies, schools, bathtubs, coral reefs, orchestras, ant colonies, Earth's climate, global economics, to name a few – are systems. Each has connecting parts, and the parts interact with one another. The sum of those interactions differs from the mere existence or behavior of the parts alone. And each system interacts with other systems – consider that human bodies are influenced by Earth's climate, and global economics may respond to what happens in schools. Each system also has a function, or purpose, that might not be obvious. A firm grasp of how systems work is crucial for understanding Earth's climate. To that end, in this chapter, we introduce systems concepts that we apply throughout the book.[1]

2.1.1. System parts and interactions

Parts of a system are usually easy to identify. They can be tangible objects (cars, coal, nests), living things (drivers, miners, penguins), or intangible (morale, expertise, information). In compiling the parts of any particular system, there are two potential pitfalls: that we will include too many non-essential parts, and that we will exclude some crucial parts. A short list of the parts of a power plant includes coal, electrical generators, and electricity. But what about power lines, cooling systems, and materials to build and maintain the plant? What about waste products such as CO_2 and particulates? What about customers, mining communities, coal trains, commodity prices? Where do we draw the boundaries between the power plant and other related systems, such as the one surrounding a coal mine? The exercise of simply listing parts reveals that systems can be complex. One challenge in studying systems is thus to include the parts that are relevant to what we seek to learn about the system and to make reasonable assumptions about the rest. Even though everything is connected to everything else, we can learn important systems behaviors by examining small sections of the whole.

Alternatively, we might ignore, or be unaware of, some crucial piece of a system, and thus be unable to explain or predict the system's behavior very well. Leaving out chlorofluorocarbons (CFCs), for example, makes the decline of stratospheric ozone difficult to explain,[2] while, in the case of accounting for colony collapse among honeybees, we have yet to learn all the relevant parts of the system.[3] We deal with the pitfall of missing parts all the time because we often have to make decisions armed with incomplete information. When outcomes surprise us, we say "If I only knew … " Political and economic systems can behave in unexpected ways, sometimes due to ignored parts, and sometimes because the nature and complexity of human behavior is inherently unpredictable. Thus, our second challenge related to system parts is to identify gaps and work to fill them when we are faced with behavior we cannot explain by what we already know. If we have high confidence that we have included all the important parts, yet we still have trouble understanding the system, it might be because we do not understand the interactions within it.

Connections and interactions among the parts keep the system functioning. These connections can be obvious tangibles: the bridge connects the two banks of the river with a surface cars can drive across; penguins interact with both food and predators. Tweaking one part of an interconnected system can produce unintended consequences. For example, burning coal dumps both CO_2 and reflective particles into the atmosphere at the same time. The CO_2 causes heating, but the reflective particles offset some of that heating. Cleaning up the particles is good for air quality, but also removes a check on CO_2 heating. Connections can also be flows of information that help the system function: an orchestra conductor directs the musicians who create the sounds of a symphony; drivers listen to traffic reports on the radio.

2.1.2. Stocks and flows

At any particular time, there is some number of cars on the bridge, coal in the hopper, and penguins in the colony. These are examples of a basic and crucial systems concept – a stock. A stock is an amount, number, or quantity of something residing in some particular place in the system at a particular time. Many stocks are tangible, but intangibles such as knowledge, political capital, or willpower can also be stocks.

Things flow in, out, and through systems. Cars drive on and off the bridge, carbon comes into the plant as coal and leaves as CO_2 gas, penguins are born and die. The *rates* at which things enter and exit are inflows and outflows. Because they are rates, inflows and outflows always have an element of time: cars driving onto the bridge *per minute,* tonnes of coal burned *per day,* penguins born per 1,000 adult penguins *per year* (Figure 2.1).

Flows influence stocks. The stock at a particular time reflects the combined history of inflow and outflow to that point. Imagine that each morning the coal plant receives 6,000 tonnes of coal and burns an equal amount each day. Then, inflow equals outflow equals 6,000 tonnes per day. The coal stock at the plant gradually decreases between deliveries, but jumps up again with the arrival of a new coal train. If the plant cuts back on power production (outflow), but keeps receiving 6,000 tonnes of coal per day (inflow), the coal stock goes up. If coal delivery gets disrupted (inflow goes to zero), the stock might gradually also go to zero if it continues to be burned at the same outflow rate. Ultimately, the outflow would also go to zero if no new coal was delivered. Stocks grow or shrink over time depending on the balance of flows. The penguin population grows if the birth rate exceeds the death rate, and shrinks if the death rate exceeds the birth rate. A bank account grows if the rate of deposits exceeds

Figure 2.1. Stocks and Flows in a System

The number of penguins in a colony is an example of a stock. Inflow and outflow are rates at which penguins are added to or subtracted from the stock. The stock at a particular time is the result of the combined history of inflow and outflow. Penguin births each year add to the stock; penguin deaths each year subtract from the stock. If inflow is greater than outflow, the stock will grow over time.

the rate of withdrawals. A leaky bucket can still be filled as long as the inflow can outpace the leakage.

Stocks influence flows, too. A large penguin colony has both more births per year (inflow) and deaths per year (outflow) than does a small colony, because the large colony has more breeding and aging adults. A large bank account accrues more interest than a small bank account because interest is based on a percentage of principal. A full bathtub drains quickly at first because the water pressure is high, then drains more slowly as it empties.

Dealing with stocks, inflows, and outflows is not intuitive. We are all humans, and humans – even the most experienced systems thinkers – are subject to pitfalls. For example, we tend to assume that the behavior of a stock over time will mimic the behavior of the most obvious inflow or outflow, and we forget or ignore other relevant flows. Consequently, our intuition about how a change in flow will affect a stock is often wrong. For a climate example, human activities add CO_2 to the atmosphere each year (inflow). As Figure 2.2 shows, the rate of inflow has increased over time, as has the amount of CO_2 in the atmosphere (stock). Inflow and stock both follow the same upward curving pattern – in other words, they are correlated. Our typical human intuition then tells us that, to stabilize or decrease the stock, we just need to stabilize or decrease the rate at which we add CO_2 each year,[4] but we would be wrong, because the stock is a record of the historical *combination* of inflow and outflow. Even if we were to stabilize CO_2 inflow at 2010 values, outflow would have to jump up to match inflow in order for the stock to stabilize. For the stock to decrease, outflow must exceed inflow. Merely decreasing or stabilizing the inflow might cause atmospheric CO_2 concentrations to rise more slowly, but the direction would still be upward, unless and until the inflow rate matches or falls below the outflow rate. If your bathtub drain is plugged, you could turn down the faucet to a mere trickle, but the tub would still fill.

When thinking about a system, ask yourself whether you have identified and considered all relevant inflows and outflows. Keep in mind that a stock might have multiple associated flows. The bridge might have sidewalks, bike lanes, and

Figure 2.2. Stocks and Flows of Atmospheric Carbon, 1900–2100: How Intuition Fails Us

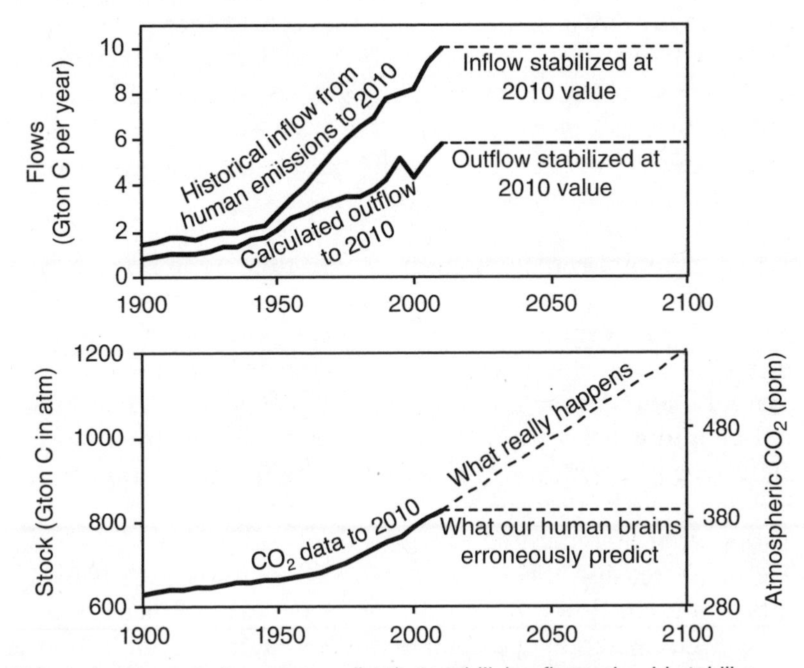

At first glance, many humans predict that stabilizing flows should stabilize a stock. But a stock stabilizes only if inflow equals outflow. If inflows and outflows of atmospheric carbon were to stabilize at 2010 values, the stock would continue to rise because inflows remain greater than outflows.

Note: Gton = gigatonne; ppm = parts per million. The left axes use units with GtonC – the mass of the carbon alone, not the whole CO_2 molecule.

Sources: Atmospheric CO_2 data (stock), 1900–60, from D.M. Etheridge, L.P. Steele, R.L. Langenfelds, R.J. Francey, J.-M. Barnola, and V.I. Morgan, "Historical CO_2 Record from the Law Dome DE08, DE08-2, and DSS Ice Cores" (Oak Ridge, TN: United States, Department of Energy, Oak Ridge National Laboratory, Carbon Dioxide Information Analysis Center, 1998), available online at http://cdiac.ornl.gov/ftp/trends/co2/lawdome.combined.dat; 1960–2010, idem, Department of Commerce, National Oceanic and Atmospheric Administration, Earth System Research Laboratory, Global Monitoring Division, from the Mauna Loa Observatory, available online at ftp://aftp.cmdl.noaa.gov/products/trends/co2/co2_annmean_mlo.txt; inflow emissions data from United States, Department of Energy, Oak Ridge National Laboratory, Carbon Dioxide Information Analysis Center, available online at http://cdiac.ornl.gov; outflow data are authors' calculations from carbon stock and inflow.

train tracks in addition to lanes for cars. Similarly, numerous flows add CO_2 to and subtract it from the atmosphere. Getting a handle on flows will allow you proceed farther in your systems thinking and to move on to the challenges of feedbacks and delays, instead of getting bogged down with inflows and outflows.

2.1.3. Feedbacks

A system might run along merrily over time, with flows in and out and with constant stocks. A glacier can stay the same size, even as it gains snow and ice from precipitation and loses them to melting. If birth and death rates approximately match, the penguin population remains constant, even though individual penguins are born and die. These situations are at steady state, or equilibrium. A simple definition of equilibrium is that inflows equal outflows.

But what if something perturbs the equilibrium, by increasing or decreasing the stock or causing an imbalance between inflow and outflow? What if the price of coal doubles? What if the penguins' food supply increases? What if a car stalls on the bridge during rush hour?

Perturbations initiate changes that "feed back" through system stocks and flows: perturbing a stock changes a flow, which changes the stock again, which changes the flow again, and so on. The loop can start with a push to either a stock or a flow. To be a feedback, the perturbation has to initiate a linked sequence of events that eventually either reinforces or counteracts the perturbation's initial push. Feedbacks that reinforce perturbations are called "amplifying" feedbacks; those that counteract perturbations are called "stabilizing" feedbacks.[5]

For a straightforward feedback example, consider what happens if you go running on a hot day. The exercise raises your body temperature. Your body reacts to this perturbation by sweating. The sweat evaporates from your skin, cooling you off, counteracting the exercise-induced temperature increase (Figure 2.3). As you cool off, your sweat production decreases again and you lose

Figure 2.3. How a Stabilizing Feedback Works

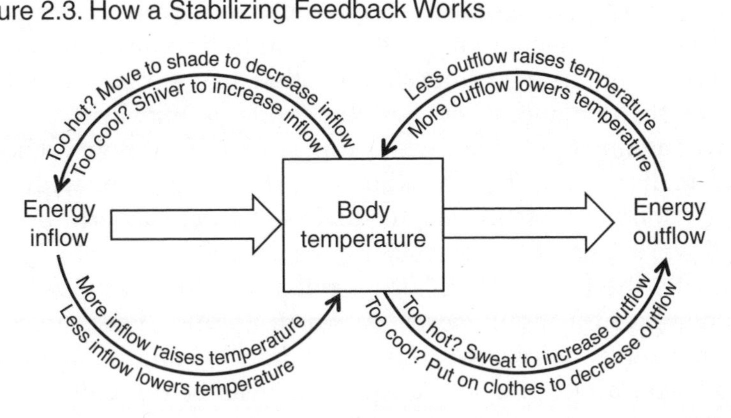

Both the left loop, involving the stock and inflow, and the right loop, involving stock and outflow, are stabilizing feedbacks that counteract hot or cold perturbations to maintain body temperature.

less heat through evaporation. This loop of responses between body temperature and sweat stabilizes your temperature in an appropriate range for a human.[6]

For a more complex but also familiar example, consider the stalled vehicle on the bridge. The stall quickly causes a traffic jam. Inflow and outflow both go to zero; the stock of cars (now stopped) on the bridge stays constant. Finding themselves stuck, some drivers call tow trucks and radio stations. A tow truck mobilizes to clear the stall. The radio stations announce the stall in their traffic reports. Drivers headed towards the bridge listen to the traffic report and some of them choose alternate routes, easing demand for the bridge. After some delay, the road is clear and traffic flows again. The "sweat" loop (action to increase the outflow) lowers your body temperature when it has been raised by exercise. Both the "tow truck" loop (action to increase the outflow) and the "traffic report" loop (action to decrease the inflow) help restore the flow of traffic.

These are examples of stabilizing feedbacks, which, as the name implies, act to stabilize a system or restore it to a stable state after a perturbation. Other examples include a thermostat that keeps your home's temperature relatively constant; a growling stomach, which inspires you to find food and alleviate your hunger; a product shortage that drives prices up, which decreases demand, alleviates the shortage, and brings prices down again; a politician who proposes a radical policy, takes a poll on its popularity, then adjusts it to be more in line with public preferences. Stabilizing feedbacks counteract perturbations and work to maintain a system in a particular state. For Earth's climate, the largest stabilizing feedback occurs because hotter things give off more energy. As Earth warms, it also gives off more energy, helping to counteract the warming. If stabilizing feedbacks are strong and act quickly, it can be difficult to move a system out of equilibrium or to a new state.

Amplifying feedbacks, in contrast, destabilize systems by reinforcing a perturbation. Consider the Greenland ice sheet. It is white and reflects a lot of solar energy back into space. As global temperature rises, ice melts. Less ice means less solar energy reflects back into space and more is absorbed by Earth's (now darker-colored) surface, helping temperature rise further, which melts more ice, and so on.

In human social systems, an advantage that accrues further advantage exemplifies an amplifying feedback. On youth sports teams, children only a few months older than their peers are slightly more mature and better coordinated physically than their younger teammates, and initially appear better at the sport.[7] They are rewarded with game time, sports camps, and coaching attention. As the years go by, the advantages keep accruing to children who were initially singled out essentially because of their age. This is why, on professional sports teams, most players have birthdays within three months after the age cutoff date for children's sports teams (for example, 1 January).

To recap, a bit of ice melt leads to further ice melt; initial athletic skill yields opportunities to improve athletic skill. Other common examples of amplifying feedbacks include microphone feedback, arms races, and viral videos. When amplifying feedbacks dominate, a system can be pushed to a new state: forests replace the ice sheet, nations newly bristle with weapons, you become famous. Amplifying feedbacks, if not checked by stabilizing feedbacks, can lead to exponential growth or decay of a stock. Amplifying feedbacks eventually do encounter limits, but often only once the system has changed to a new, sometimes irreversible, state. The gravitational attraction of newly formed Jupiter sucked in nearby material in the early solar system, making Jupiter bigger, increasing its gravitational attraction, making it bigger, and so on. But once Jupiter swept up nearly everything near it in the solar system, it stopped growing (Figure 2.4). An ice sheet cannot melt any further once it is gone. To keep Earth's climate system within a "stable" range, humans ought to avoid crossing thresholds that initiate strong amplifying feedbacks.[8]

Systems can have both amplifying and stabilizing feedbacks. Biological food webs are good examples. Consider penguins. A perturbation that increases penguin food (fish, shrimp) might mean more chicks mature to breeding age. More breeding penguins means more chicks in the next generation, and the penguin population grows. This is an amplifying feedback. The feedback loop might end when the penguin population reaches a new, higher number that approximately matches the new food supply. Or, the penguins might outstrip their food supply and starve, decreasing their population (a stabilizing feedback). With

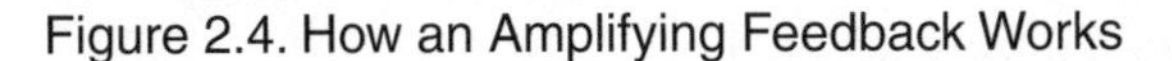

Figure 2.4. How an Amplifying Feedback Works

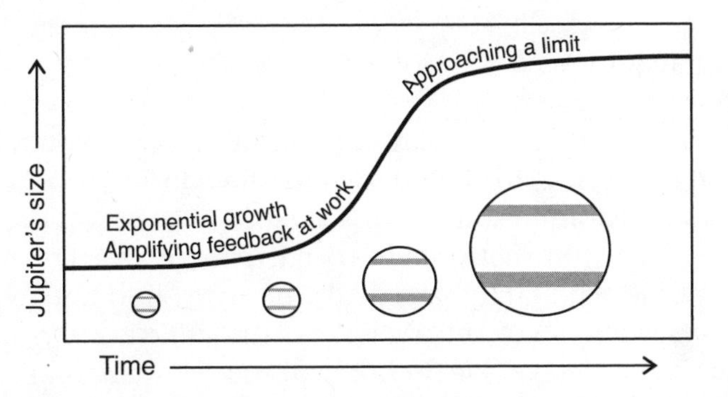

Exponential growth from an amplifying feedback ends when some physical limit is reached, as in the formation and growth of Jupiter.

less penguin pressure, fish and shrimp populations grow again. Enter leopard seals, which eat penguins. With more penguins around for food, the leopard seal population grows, then falls again, following, but lagging behind, the penguin population. The stocks (populations) in biological systems can oscillate up and down, as various feedbacks operate. Humans are currently on a population upswing, following our discovery and use of a big new energy source: fossil fuels. Like those of penguins and leopard seals, our population will not grow forever.

To summarize, feedbacks are more than cause-and-effect chains. To be a feedback, the effect has to turn around and kick the cause. Amplifying feedbacks destabilize systems; stabilizing feedbacks help balance them. Amplifying feedbacks push a system towards a new state; stabilizing feedbacks resist change. In the climate system many feedbacks, both stabilizing and amplifying, are in operation. A useful way to think about climate feedbacks is to ask: does this part of the system respond to changes in Earth's temperature and, in return, influence temperature?[9] If the answer is yes, you are dealing with a climate feedback. Common climate feedbacks involve water vapor, clouds, ice, carbon, and infrared radiation. Some feedbacks in Earth's climate system are well known, others less so. Much effort continues to be expended in estimating the magnitudes and directions of important feedbacks to better constrain the probability of future climate scenarios.

2.1.4. Lags

Systems have inherent delays. It takes time for ice on a lake to melt in the spring, for your lunch to move through your digestive system, or for a ship to turn or stop.

A lag is the time between the occurrence of an action and the effect of the action. Humans deal well with short and expected lag times: a phone conversation flows smoothly even with minor transmission delays. We do not deal so well with long and unexpected lags. In some bathroom shower systems, the hot and cold water get mixed at a point in the piping several feet away from the showerhead. When you change the hot/cold mixture, it takes time for the new mixture to reach you. If the water is too hot, you add cold, but since the temperature does not respond quickly (because the water is still in the pipe on its way to the showerhead), you might add even more cold because you want a faster response. When the new mixture arrives, it quickly gets too cold. You turn down the cold and crank up the hot. But the water coming out of the showerhead is still responding to your previous "cold" command, so you crank up the hot some more. Soon, you are scorching. Even though you are seeking an equilibrium, and use "stabilizing" actions, delays in the system keep the shower water temperature oscillating around the desired value.

Lags are not intuitive. It is easy to ignore slow-acting responses. In the climate system, human additions of CO_2 to the atmosphere have induced an energy imbalance. The surface ocean absorbs some of the extra energy, mixes it downward, and heats the water below. But this mixing takes time, and because of the delay, Earth's surface temperature has yet to respond fully to the human-induced energy imbalance. Even if we could hold atmospheric CO_2 at today's values, surface temperature will continue to rise as the ocean slowly equilibrates, just as a pot of water on the stove gradually heats, even if the energy from the burner is held constant. Equilibration will take even longer under a scenario of increasing energy imbalance.

2.1.5. Function or purpose

The function or purpose of a system is determined by its actual behavior. The stated purpose of a bridge expansion project might be to help people travel from A to B faster than they could before. But infrastructure expansion sometimes has the unintended consequence of attracting new development and more traffic, and pretty soon the bigger bridge is just as clogged as the old one. The stated purpose of a government might be to provide beneficial services to a community, but what actually happens under a corrupt government could be quite different. These are simplistic examples, but the point is that, in the words of Donella Meadows, "the best way to deduce [a] system's purpose is to watch for a while to see how the system behaves."[10]

Systems can surprise us. They are self-organizing, self-perpetuating, and they evolve. They can have thresholds beyond which their behavior changes. Try to push a boulder over a cliff with your finger, and you will likely fail. But once you and your friends get the boulder moving, it (and anything it encounters) heads towards an entirely new state. Even well-known nonlinear behavior, such as exponential growth, often surprises us. The king who lost at chess and had agreed to pay the victor one grain of rice on the first square, two on the second square, and continuing to double for each square of the chessboard realized quickly – but too late – his imminent bankruptcy.[11] Systems thinking takes a lot of practice.

2.2. Earth's climate system: The parts and interconnections

2.2.1. Atmosphere, hydrosphere, biosphere, geosphere, and anthroposphere

Earth's climate system is complicated. Sometimes, it helps to examine the parts separately. At other times, it helps to represent the whole system at once and see how it behaves. Consider the five big theatrical stages in the climate system: the atmosphere, hydrosphere, biosphere, geosphere, and anthroposphere. Sometimes, dividing the world like this provides useful boundaries for thinking, because important subsystems – for example, social organizations, chemical reactions in the atmosphere or ocean, population dynamics – operate on each of the stages. But keep in mind that some characters – such as clouds, farms, volcanic ash – clearly act on more than one stage, and that the five spheres are so interconnected that you often need to consider more than one at a time.

2.2.1.1. *The atmosphere*

Without the atmosphere – that thin veneer of gases above Earth's surface – we would not even be here. Ozone in the stratosphere (about 10–50 km above the surface) absorbs incoming high-energy ultraviolet radiation from the Sun, providing biological organisms with some protection. Greenhouse gases, mostly in the troposphere (from the surface to about 10 km up), absorb and re-emit long-wave radiation coming from Earth's surface, slow the passage of energy from the surface back into outer space, and warm the planet. Although these two functions are probably of greatest benefit to life on Earth, the atmosphere also (1) contains the oxygen we breathe; (2) hosts clouds, which contribute to the greenhouse effect, reflect incoming solar radiation, and provide a means to transfer water from oceans to land; (3) hosts tiny particles and droplets that reflect and absorb energy; (4) blows around to redistribute energy, keeping the poles warmer and the tropics cooler than they otherwise would be; and (5) drives surface ocean currents, which are intimately linked to marine productivity, on which humans rely for food, and which, in the long term, provide the material to make fossil fuels, our current energy source.

The atmosphere is made up of gases and tiny particles and droplets. Over the lifetime of the planet, the composition of the atmosphere has changed, but today the gases are mostly N_2 (nitrogen, about 78 percent of the atmosphere), O_2 (oxygen, about 21 percent), and Ar (argon, about 1 percent). As well, there are a number of trace gases in small amounts, the most important for the greenhouse effect being H_2O (water), CO_2 (carbon dioxide), CH_4 (methane), and N_2O (nitrous oxide). Famous O_3 (ozone) holds down two jobs: it absorbs harmful short-wave (ultraviolet) radiation from the Sun, as noted, and it acts as a greenhouse gas by absorbing long-wave radiation from Earth's surface. It is important to differentiate between these two separate roles. All the gases continually bounce around, colliding with one another, so they are well mixed.

The particles and droplets in the atmosphere are called aerosols. They are so tiny that gravity does not pull them immediately down to Earth's surface. Some – sulfuric acid droplets, for example – are excellent reflectors of solar energy. Others – such as black carbon (soot) – absorb solar radiation and contribute to warming. Aerosols can also have indirect effects. For example, they provide surfaces for water droplets to form on, which affects cloud formation and potentially changes the reflective properties of clouds. The net effect of aerosols is cooling: they decrease the amount of solar energy available for the lower atmosphere and Earth's surface to absorb.

Living on the planet's surface is like living at the bottom of a giant swimming pool. In the troposphere – the lower part of the atmosphere up to about 10 km – air moves in every direction, mixing the air molecules. Gravity keeps most of the gases and particles in the atmosphere close to the surface, so the air gets thinner higher up, although the relative proportions of gases such as N_2, O_2, and CO_2 are about the same everywhere. Air pressure also diminishes as you go from sea level up to higher altitudes. Choose a location and measure air temperature upward through the atmosphere. At a typical spot, with water vapor in the air, temperature cools about 7°C per km of altitude. It is therefore about 70°C colder at the top of the troposphere than at sea level. That is why airplanes flying 10 km up must heat and pressurize their cabins to keep us comfortable, and why mountain climbers wear warm clothing and supplement their air with bottled oxygen if they are high enough. The steepness of this temperature gradient – the lapse rate – varies with time and location. It both responds to global temperature change and influences global temperature – a climate feedback we will visit in Chapter 5.

Above the troposphere is the stratosphere, where the temperature gradient reverses direction to increase with increasing altitude. Warmer air that sits above colder air is quite stable, so the reversed temperature gradient in the stratosphere is a big reason the troposphere and stratosphere do not mix together much. And because they do not mix, they have different compositions. For example, ozone is present in higher concentrations in the stratosphere than in the troposphere, and aerosols that get into the stratosphere – from, say, a powerful volcanic eruption – can stay up there much longer (a few years) than they can in the troposphere (a few weeks), where they get rained out. There is little water vapor in the stratosphere, though what little there is continues its role as a greenhouse gas.[12]

In summary, the atmosphere is where much of the important climate action happens. Its composition has changed over time and is changing now. Flows of carbon, water, aerosols, energy, and so on into and out of the atmosphere are crucial to the behavior of the climate system, because flows determine the stocks of each of these key ingredients. Atmospheric feedbacks are typically fast acting, and some atmospheric processes have very short response times. Others, because of interactions with the oceans or biosphere, take longer to respond. If we stopped burning coal today, the associated aerosols would fall out of the atmosphere quickly, but the resulting CO_2 would stick around for a long time.

2.2.1.2. The hydrosphere

Heavy rainfall, flooding, drought, ice storms, and hurricanes remind us, in a very localized and immediate way, of the power of water in all its forms. Over twenty million people's lives were disrupted by extreme flooding in Pakistan in 2010.[13] Hurricane Katrina in 2005 caused an estimated US$108 billion in damages.[14] By the time you read this book, many other large-scale, water-related disasters will have occurred. Overall planetary heating increases the probability of severe weather, often related to the water cycle.[15] And although extreme events garner immediate attention, sea-level rise is a slower-acting but more widespread climate problem, both for those who live near the coast and those occupying higher ground, where coastal dwellers might seek to move.

The hydrosphere plays at least four important roles in the climate system: the ocean absorbs, transports, and releases energy; water vapor acts as a greenhouse gas; the ocean uptakes, transports, and releases CO_2; and water and ice reflect solar energy differently.

Why do we flock to lakes, rivers, and oceans during hot weather? Coastal dwellers know firsthand the benefits of proximity to open water – milder winters and milder summers. Water has a high heat capacity, meaning that it takes a lot of energy to change its temperature. To heat 1 gram of liquid fresh water by 1 kelvin (K), takes about 4.18 joules (J) of energy (a little less for saltwater). When a gram of water cools off by 1 K, it releases 4.18 J. But who has an intuitive sense of what 4.18 J/g*K means? Think of it this way: heating a teapot's worth of water (about 1,000 g) by 100 K from the freezing point to the boiling point takes about the amount of energy in an apple, while 100 billion Hiroshima-like atomic bombs[16] would raise the temperature of the entire ocean by 1°C (1 K). On a global scale, the vast volume of ocean water moderates temperature swings by absorbing and releasing energy. Estimates of ocean heat content show that, over the past several decades, the ocean has absorbed enormous quantities of energy, the equivalent of about 2.4 Hiroshima-like bombs every second,[17] slowing the rate of global temperature rise both because water has a high heat capacity and because the ocean mixes slowly.

Water vapor, a greenhouse gas, is involved in one of the most important feedbacks in the climate system. Water responds to temperature changes quickly by shifting among vapor, liquid, and solid phases. At colder temperatures, less water is present in the vapor phase, more in liquid and/or solid phases; at warmer temperatures, more is in the vapor phase. When you "see your breath" on a cold day, what you are actually seeing is the water vapor from your warm breath quickly condensing in the cold air. An increase in atmospheric temperature causes more water molecules to move to the vapor phase. As a greenhouse gas, that water vapor further increases the temperature, amplifying the initial warming. A stabilizing feedback keeps the water vapor in check – when the atmosphere reaches saturation, it rains. In a warming world, with more total water vapor in the atmosphere, that can mean a lot of rain.

Every year, humans convert some amount of fossil fuel (coal, oil, natural gas) to energy and, as a product of combustion, emit

CO_2 into the atmosphere (an inflow). In 2011 globally, humans emitted about 9 gigatonnes of carbon[18] (GtonC) by burning fossil fuels.[19] Adding cement making and changes in land use brings total carbon emissions that year to about 10.4 GtonC. If all those emissions had stayed in the atmosphere, atmospheric CO_2 concentration (stock) would have increased about 4.9 ppm that year.[20] Instead, the stock increased by only about 2 ppm.[21] The oceans absorbed (an outflow) about one-quarter of the 10.4 Gton, so they slow, but do not totally compensate for, human-related flows of CO_2 into the atmosphere. In addition, this oceanic uptake of CO_2 has side effects: the CO_2 increases the acidity of oceans, with ramifications for marine biota and, as atmospheric CO_2 increases further, the oceans' chemical capacity to continue to remove CO_2 from the atmosphere is expected to slow down, making it a less effective outflow.

The reflectivity of water in its various forms is integral to important climate feedbacks. The liquid oceans, which cover a large proportion of Earth's surface, are fairly dark and absorb about 95 percent of the incoming solar radiation that reaches them. But glaciers, ice sheets, sea ice, and water droplets and ice crystals in clouds are all highly reflective, and send solar radiation directly back into space. Fairly straightforward amplifying feedbacks with ice growth and melt (section 2.1.3.) change the overall reflectivity of Earth's surface. Cloud reflectivity feedbacks are more complicated, but, to follow one line of reasoning, if warmer temperatures promote more cloud formation, those clouds would reflect more incoming solar energy and cause cooling. This would counteract the initial warming – a stabilizing feedback. On the other hand, in a warmer world, it is likely that larger areas in and near the subtropics will have less cloud cover, decreasing overall global cloud reflectivity and amplifying the warming.[22]

In summary, the vast oceans slow the rate of global temperature change and take some of the excess CO_2 out of the atmosphere. Water vapor is a key greenhouse gas. Ice and clouds mostly reflect incoming solar energy, and liquid water mostly absorbs it. And, of course, water in lakes, rivers, oceans, and even ice provide living space for a vast array of organisms that are themselves part of the climate system.

2.2.1.3. The biosphere

The biosphere plays two fundamental roles in Earth's climate. First, living things influence stocks and flows of materials such as carbon, water, and nitrogen. Second, the type and extent of vegetation cover on Earth's surface influence reflectivity and therefore how much solar radiation is absorbed by the surface or directly reflected back into space.

Earth's plants and animals are all carbon-based life forms, and carbon flows in and out of the "living biosphere" stock. When plants photosynthesize, carbon flows from the atmosphere into the biosphere, and when an animal eats a plant, carbon flows from one part of the biosphere to another. Carbon flows also from the biosphere back into the atmosphere when plants and animals respire or decay or combust. A very small amount of the carbon from the biosphere gets buried and goes to the geosphere each year (Figure 2.5).

On a one-year time scale, seasonal plant growth and decay are the most important processes influencing atmospheric CO_2. From spring to autumn, when the rate of photosynthesis (outflow from atmosphere) exceeds the rate of respiration and decay (inflow to atmosphere), the stock of CO_2 in the atmosphere goes down. The stock reaches a low in autumn, when the respiration rate catches up with, then exceeds the photosynthetic rate and carbon flows back to the atmosphere throughout the winter. Since seasons are opposite in the northern and southern hemispheres, the seasonal variations in atmospheric CO_2 stock are also opposite.

As long as the annual average stock of carbon in the living biosphere stays approximately constant from year to year, these seasonal ups and downs balance out. This is why your breathing does not have any effect on the current long-term rise in atmospheric CO_2. You breathe out carbon that a plant (your food) removed from the atmosphere probably within the previous year.[23] If, however, the stock of carbon in the living biosphere goes up or down, the atmosphere will feel the effects.

Figure 2.5. The Carbon Cycle

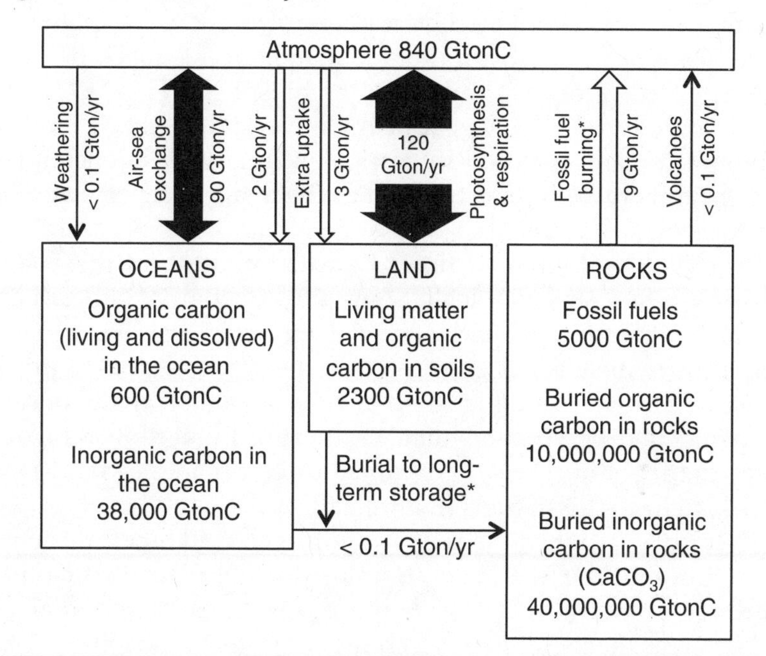

The figure shows a simplified carbon cycle, with perturbation from fossil fuel burning by humans. Flows related to human activities are open arrows; natural flows are solid. With natural processes, about 210 GtonC per year flows back and forth between the atmosphere and the land and oceans. Since the flows balance, they do not alter the stocks. Fossil fuel burning by humans introduces a new flow from the buried organic carbon stock to the atmosphere. About half of the excess flow is taken up by biology on land (3 GtonC per year) and by the oceans (2 GtonC per year). The remainder stays in the atmosphere and increases the stock of atmospheric carbon. Asterisks indicate flows with potential for mitigation efforts – that is, decreasing fossil fuel burning and increasing long-term burial. For simplicity, the long-term flows (weathering, burial to long-term storage, and volcanic outgassing) are all shown as the same small value (less than 0.1 GtonC per year), and natural weathering of carbon-rich rocks is not shown.

A massive insect infestation that kills a lot of trees depletes the stock of carbon stored in a forest, returning much of it to the atmosphere.[24] In contrast, the abandonment of much farmland in eastern North America and the subsequent natural reforestation there has taken carbon from the atmosphere and added it to the living biosphere stock.[25] Some evidence suggests that plant growth is stimulated by higher atmospheric CO_2 concentrations, which would promote a stabilizing feedback if it led to increased storage of carbon in the biosphere. This "CO_2 fertilization effect," however, appears to cause little change in the amount of carbon stored in plants, and plants also encounter other limits to their growth, such as the availability of water or nitrogen.[26]

The nice neat circle of flows between plants and the atmosphere each year has a leak. Some tiny fraction of the carbon in plants gets buried and does not return to the atmosphere right away. Instead, it goes into another carbon stock: buried organic carbon. "Organic" here just means matter made of carbon and hydrogen atoms that was part of living things before they died, got buried, and became buried organic carbon. This stock includes fossil fuels (Figure 2.5).

Not only carbon, but also water, nitrogen, and other nutrients flow through the biosphere. Rainforests fuel their own clouds through evapotranspiration, helping perpetuate high rainfall in those areas. Bacteria turn N_2, abundant in the atmosphere, into forms of nitrogen plants can use. Other bacteria help decompose organic matter, returning the nitrogen to the atmosphere. Since the early 1900s, humans have turned N_2 artificially into nitrogen fertilizers for agriculture, causing an imbalance in the nitrogen cycle. These fertilizers fuel soil microbes that produce N_2O, a greenhouse gas, at a faster rate than would otherwise happen.[27]

A secondary role for the biosphere, in the context of climate, is that vegetation cover contributes to how much solar energy is absorbed versus how much is reflected back into space. Dark forests absorb more energy than light-colored grasslands and deserts. Expansion of shrubs and boreal forest northward darkens the paler tundra.[28] Desert expansion and replacement of forests with fields lightens other areas. These changes have small effects on Earth's energy budget.

In summary, the most important climate roles the biosphere plays involve reflection, and carbon, water, and nitrogen cycling, but it also plays other climate-related roles. Marine organisms produce compounds that, when entrained in the atmosphere, might help clouds form.[29] Plant roots slow erosion by holding soil in place, but also widen cracks in rocks to help break them apart. Plants pump CO_2 down into the soil and help in weathering. Forest fires add CO_2 to the atmosphere. The list of biosphere credits is long.

2.2.1.4. The geosphere

The geosphere encompasses all the geologic materials on Earth: the rocks, soils, marine sediments, sand, mud, fossil fuels, and so on. Large-scale geologic conditions today – the positions of continents, locations of volcanoes, altitudes of large mountain ranges, widths and depths of the ocean basins, locations of fossil fuel deposits – result from slow geologic movements up to this point in Earth's history. For thinking on human time scales, these features can be considered stable: we can neither change them quickly nor do we need to worry much about them. We do, however, need to recognize that this particular scenario is the basic reality of our planet today. So, when we construct climate models, we need to include today's real distribution of land masses and oceans at whatever level of detail needed for the particular questions we want to address.

Three components of the geosphere are particularly important for climate: volcanoes, weathering processes, and fossil carbon.

Explosive volcanoes occasionally make the news when they spew ash particles into the atmosphere, disrupting air travel. The ash falls to the ground within weeks, but volcanic gases can influence climate for up to a few years. In particular, sulfur dioxide gas combines with water and oxygen to form sulfuric acid droplets – aerosols. If a volcano is powerful enough to blast these into the stable stratosphere, they will settle out over several years; meanwhile, high-altitude winds distribute them around the globe. The sulfur aerosols reflect incoming solar energy, decreasing the amount of energy Earth absorbs, cooling the planet. A single large volcanic eruption can decrease global average surface temperatures by 0.1 to 0.2°C for a year or two.[30] For example, the eruption of Mount Tambora in Indonesia in

1815 resulted in the "year without a summer" and widespread crop failures. The eruptions of Krakatoa, also in Indonesia, in 1883, Pinatubo, in the Philippines, in 1991, and others were powerful enough to decrease incoming solar energy measurably. On human time scales, volcanoes are temporary cooling devices.

We cannot control volcanoes, but they are nonetheless important for climate mitigation and adaptation. First, volcanoes provide natural laboratories in which we can study how Earth's climate responds to energy perturbations. We can observe and measure the decrease in energy absorbed due to volcanic aerosols and, at the same time, measure their effects on global temperature. Modeling these short-term changes can help us estimate how climate will respond to energy perturbations from other sources. Second, natural processes can provide ideas for mitigation solutions. Volcanoes have a temporary cooling effect. Could we artificially imitate what volcanoes do to counteract the heating from human-sourced atmospheric CO_2? What would it take, and what might be the consequences? We explore this question in Chapter 4.

On short time scales, volcanoes reduce global temperatures, but over millions of years volcanoes can heat the planet. Volcanoes release CO_2, although, compared with human emissions, it is a tiny amount each year.[31] But imagine if Earth had a particularly high rate of volcanism for a million years. CO_2 inflow to the atmosphere would go up, initiating an important feedback. As atmospheric CO_2 increased, temperature would increase, more water would evaporate, and rainfall would go up. Moreover, the rainwater would be acidic because CO_2 and water combine in the atmosphere to form carbonic acid. The acidic rain would fall on rocks and chemically break them down. Higher atmospheric CO_2 thus would lead to higher rates of rock weathering. In turn, rock weathering would take CO_2 out of the atmosphere. This is a stabilizing feedback that operates on very long geologic time scales. In the future, rock weathering will bring atmospheric CO_2 back down long after our CO_2 emissions experiment ends. Although the response of rock weathering to volcanic emissions of CO_2 is a long-term feedback, it too inspires possible mitigation ideas. Can we artificially increase rates of rock weathering to remove CO_2 from the atmosphere?

Finally, there is fossil carbon, which is central to this book. Fossil fuels form when organic carbon leaked from the biosphere gets squeezed at high temperatures and pressures, then accumulates in places from which humans then extract it. In reality, much more organic carbon is buried in tiny, dispersed bits than is available in exploitable fossil fuel stocks. In a balanced system, buried carbon cycles slowly to the surface, then erodes and oxidizes at about the same rate as new burial. But imbalances happen. For example, during the Carboniferous Period, about 300–350 million years ago, the rate of organic carbon burial exceeded the return flow, the atmosphere lost a lot of carbon, and Earth got cold. Much of the coal we burn formed during that period. Today, the imbalance is opposite: humans have greatly increased the outflow of carbon from the fossil fuel stock (Figure 2.5), and although the biosphere, both on land and in the ocean, absorbs some of the excess, it cannot keep up. There is no fast-acting mechanism to counter today's flow of carbon from fossil fuels to the atmosphere. Instead, we have increased the probability of amplifying feedbacks whereby greenhouse-induced warming helps release more ancient buried organic carbon currently trapped by permafrost and frozen ocean sediments,[32] and we do not have a direct way to prevent that from happening.

2.2.1.5. *The anthroposphere*

Humans are an incredibly successful species if we measure success by population, versatility, and impact. We have learned how to harness an ever-changing combination of food and non-food energy supplies. Wood, dung, draft animals, charcoal, whale oil, wind, coal, oil, natural gas, solar energy, nuclear energy, hydrogen fuel cells, and others have all done work for us. Our discovery and expanding use of fossil fuels have generated exponential growth in human numbers and made life for some of us quite comfortable.

Signs of humans are everywhere. The history of our expansion across the globe is written in extinctions of vulnerable species shortly after our arrival in new places. Early agricultural efforts and land clearing left imprints on the geologic record and on surface features that are observable today. Our actions now move more rocks and soil every year than all natural processes combined. If you fly over Western Europe, for example, on a clear day, you will be hard pressed to spot land surfaces untouched by human activity.

Humans deeply influence the composition of the atmosphere as well. The most obvious example is our rapid addition of CO_2 from burning fossil fuels. Burning long-buried fossil carbon is fundamentally different than burning, say, corncobs, which took their carbon out of the present-day atmosphere only a few months ago. With corncobs, CO_2 inflow and outflow are balanced; with fossil carbon, inflow greatly exceeds outflow.

CO_2 is of primary importance, but it is not the only atmospheric constituent that humans alter. Agriculture adds N_2O, a greenhouse gas. Industrial processes add human-made chemicals, such as CFCs, which help deplete ozone in the stratosphere. We add soot and other particulate matter that reflects and absorbs energy. To some degree, we have managed to reduce our emissions (inflows) into the atmosphere: scrubbers, for example, prevent the emission of some particles and gases from the burning of coal; as a result of the Montreal Protocol, the production of CFCs has largely been phased out (although replacements have created other problems); more efficient technology is yielding more usable energy per unit of fossil fuel burned. But we have just barely begun efforts to increase outflows – meaning, for example, sucking large quantities of CO_2 out of the atmosphere and putting it someplace it will stay for a long time. Both the decrease of inflows and the increase of outflows are potential areas for mitigation.

Large-scale changes in the world's oceans are also linked to human activities. Over time, we are changing the marine foods we eat as, one after another, the easy-to-catch, tasty organisms decline in number – indeed, examples abound of human overfishing of particular marine resources.[33] More subtle effects include large-scale ocean acidification as the oceans absorb some of the extra atmospheric CO_2; rising water temperatures, which affect the geographic range and survival of temperature-sensitive species; and rising sea levels, which alter the nature and space available to near-shore marine organisms.

Human effects on the biosphere are numerous. We have helped certain desirable (to us) plant and animal species expand their range[34] and evolve through our hybridization efforts; vaccines have eradicated some nasty diseases, such as smallpox. Knowingly and unknowingly, we have driven some species to extinction. When we introduce exotic species to new places, intentionally or unintentionally, some thrive and alter the ecosystem structure of their new environment to the detriment of already-established species. Our production and release of chemicals, such as certain pesticides, diminish the numbers of both targeted and untargeted species. Our agricultural effluents fuel algal blooms that otherwise are less likely to occur. Our management of forests has increased vulnerability to fire and insect outbreaks. Our development and use of antibiotics has saved lives, but it has also bred superbugs. We are not immune to our own actions.

In summary, the range of humans is now the entire surface of the planet. Our effects extend across the surface, up through the atmosphere, and down through the deep oceans. Humans are, of course, also part of the biosphere, but addressing the challenge of climate change involves human choices, so, for the purposes of this book, we will indulge in some species exceptionalism and consider ourselves a separate category.

2.2.2. The ins and outs of Earth's energy budget

The three basic controls on Earth's climate are solar energy, Earth's reflectivity, and the greenhouse effect. Each contributes to how much energy enters and leaves the planetary climate system. In this short introduction, we give the complete picture of Earth's energy budget. In Chapters 3, 4, and 5, we take the budget apart and focus on each of the big three controls in turn.

2.2.2.1. Does what comes in go out?

First, spend a few minutes with Figure 2.6, which illustrates Earth's energy budget.[35] The arrows represent flows of energy going different places. The labels reveal the processes by which the energy is transferred around. The numbers are in watts per meter squared (W/m^2). The unit "watts" is the same as joules per second (J/s), so W/m^2 is a measure of energy going through a given surface every second. To illustrate the meaning of W/m^2,

Figure 2.6. Earth's Energy Budget

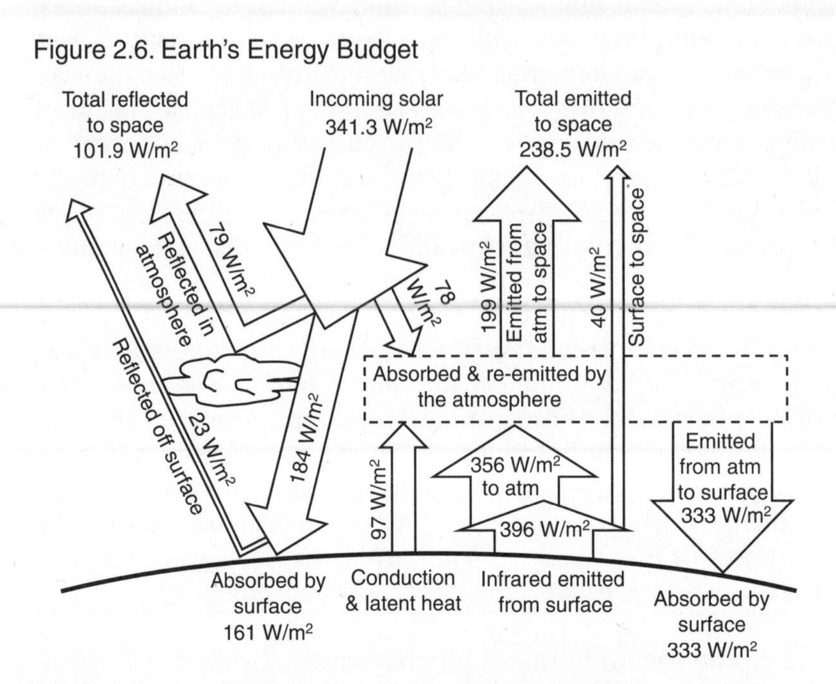

The figure shows Earth's global annual radiation budget based on data from 2000 to 2004. Width of arrows correlates to magnitude of energy flow. The reflection arrows on the left involving short-wave, solar radiation are the subject of chapter 4. The absorption and emission arrows on the right involving long-wave, infrared radiation are the subject of chapter 5.

Source: Modified from Kevin E. Trenberth, John T. Fasullo, and Jeffrey Kiehl, "Earth's Global Energy Budget," *Bulletin of the American Meteorological Society* 90, no. 3 (2009): 311–23, figure 1. ©American Meteorological Society; used with permission.

imagine, for example, a picnic blanket that measures one meter square, spread on the ground. Now imagine the entire surface of Earth covered with such picnic blankets, edge to edge. The blankets near the equator receive and absorb more than the blankets at the poles. The blankets absorb solar energy only when they are on the daytime side of the Earth, and the blankets in the summer hemisphere absorb more than those in the winter hemisphere. Now, if we average over time (over at least one year) and over the full extent of Earth's surface, the average picnic blanket absorbs 161 joules of energy every second ($161\,W/m^2$). At your picnic, you would get that much energy eating about one and a half apples per hour.

Now, look at the different processes in the figure and their relative magnitudes. The Sun sends in about $341\,W/m^2$ – in Chapter 3, we delve into the details of this number and how it changes with time. Notice that about 30 percent (about $102\,W/m^2$) of the incoming solar radiation reflects off Earth's surface and off clouds, dust, and aerosols in the atmosphere, and heads immediately back into space – reflection is the topic of Chapter 4. Imagine ways in which this 30 percent could change, or actively be changed by human activities. Notice the magnitudes of energy involved with greenhouse gases – the subject of Chapter 5. These gases absorb infrared radiation emitted from the surface, then re-emit it. Some of this radiation eventually goes back towards Earth's surface, and some exits back into space.

Notice that energy can leave Earth's surface in three ways: through radiation, conduction, and latent heat transfer.

Infrared radiation emitted from the surface has the largest magnitude ($396\,W/m^2$). Anything with a temperature above absolute zero can radiate energy, and hotter objects radiate more energy. Earth radiates energy in accordance with its temperature. Surface radiation is the big path, but energy also leaves the surface through conduction and latent heat transfer.

Conduction is energy transfer between molecules through contact or collision. For example, a hot molecule vibrates and makes the molecules next to it vibrate, too, transferring energy to the adjacent molecules, or gas molecules in the atmosphere collide, transferring energy. When your cup of coffee warms your hands, conduction is the main reason. Conduction of energy from Earth's surface can warm the lower atmosphere and initiate convective thermals – rising pockets of warm air that transport the energy higher into the atmosphere.

Latent heat transfer happens when water molecules evaporate and leave the surface in vapor form, then condense in the atmosphere. To evaporate water takes energy, and that energy comes from Earth's surface, just as the energy to evaporate your sweat comes from your body itself. When the water later condenses in the atmosphere, that vapor-to-liquid transformation releases energy. Evaporation and condensation together transfer energy from Earth's surface to the atmosphere. The same process is at work with evapotranspiration, where the water vapor leaves Earth's surface via plants, rather than evaporating from water bodies or from soil.

Some of the surface radiation that Earth emits (about $40\,W/m^2$) escapes directly into space through the "atmospheric window" – infrared wavelengths that happen to be free from interference by greenhouse gases. But much of the radiation is absorbed by greenhouse gases. The greenhouse gases constantly collide with neighboring molecules in the air – mostly N_2 and O_2 – and these collisions transfer energy, warming the atmosphere. Greenhouse gases also re-emit some of the energy by radiating it in random directions. The effect of this random re-emission is that some of the energy goes back to Earth's surface to be reabsorbed and some of it goes to the top of the atmosphere and escapes into space. Along the way, the energy might be absorbed and re-emitted by lots of different individual greenhouse gas molecules. The more greenhouse gas molecules in the atmosphere, the "slower" the passage of energy from Earth's surface to space and the higher the "stock" of energy in the atmosphere.

It is important to note that Figure 2.6 focuses on flows – there are no numbers for energy stocks in the diagram. From a systems perspective, the question of interest here is whether Earth's energy system is in balance – that is, whether inflows match outflows. If they do not, how big is the mismatch and which flow is bigger? How will the energy imbalance alter Earth's stock of energy and, therefore, climate? For the period from 2000 to 2004, Earth's energy imbalance was about 0.9 W/m^2. At the top of the atmosphere, Earth received (inflow) about 0.9 W/m^2 more than it sent back into space (outflow). Excess energy raises Earth's temperature. As long as imbalance continues, temperature will continue to rise.

2.2.2.2. Climate sensitivity: How much bang for your buck?

How much temperature change is caused by a sustained energy imbalance of 0.9 W/m^2? During a long summer barbecue, how drunk would you get from drinking one bottle of beer per hour? What about two or three bottles per hour? How drunk would your friends get? How fast? Once you start drinking, the alcohol in your system gradually builds up until you reach your equilibrium drunkenness. At equilibrium, your body gets rid of alcohol at the same rate that you drink it. Drunk and sober are two equilibrium states along the spectrum of possibilities from stone cold sober through faint buzz, intoxicated, alcohol poisoning, or worse. Each individual's characteristics – weight, gender, liver health, genetics, food consumed recently, medications – contribute to one's reaction to a perturbation in blood alcohol. We could call this "drunkenness sensitivity." A ninety-eight-pound, weak-livered person drinking all afternoon on an empty stomach would have high sensitivity. A two-hundred-pound healthy-livered person drinking while eating would have lower sensitivity.

In terms of Earth, how sensitive is the climate system to perturbations in energy? How much additional temperature rise would occur for a 0.9 W/m^2 imbalance? How long would it take for energy outflow to catch up with inflow, and when it did, what would be Earth's equilibrium new temperature? Does Earth have high or low sensitivity?

"Climate sensitivity" is broadly defined as the surface air temperature change resulting from an energy imbalance in the system, once the system has adjusted and reached a new equilibrium state. Thus, climate sensitivity is a key piece of information in climate science, particularly in making forecasts of the future. To respond to an imbalance and to establish a new equilibrium temperature takes time, just as you do not get immediately drunk. Earth has partially, but not yet fully, responded to the energy imbalance induced by humans' increased emissions of greenhouse gases – if it had, we would be back to balance, with inflow equaling outflow at a new, stable temperature.

Scientists approach the problem of estimating climate sensitivity from numerous independent angles. The actual estimates have not changed much in thirty years, but as more independent lines of evidence converge on similar results, confidence in the estimates increases. The answers cluster around a central value of about 3°C for every additional 4 W/m^2 of energy.[36] Climate sensitivity is often expressed this way because 4 W/m^2 is about the additional energy we get from doubling atmospheric CO_2 concentration, which is a useful and informative case to study for possible future scenarios. Think of it as about 0.75°C for every 1 W/m^2. This temperature response estimate includes the effects of fast-acting feedbacks in the climate system, those involving clouds, water vapor, aerosols, and sea ice. All these respond quickly to temperature change *and* feed back quickly to influence temperature. The estimate does not include slow-acting responses such as ice-sheet growth and decay or deep ocean mixing.

No one can say with absolute certainty how warm Earth will get from each extra W/m^2, so climate sensitivity, like alcohol sensitivity, has a distribution of estimates. Values between 1.5°C and 4.5°C per 4 W/m^2 appear most probable.[37] Really low values can be eliminated because the behavior of CO_2 in the atmosphere is well established: greenhouse warming that considers only the absorbing and re-emitting of radiation by CO_2 – ignoring known fast-acting feedbacks – yields 1.2°C per 4 W/m^2. Very low values for climate sensitivity are therefore mere wishful thinking. On the high end, evidence from the last ice age implies that climate sensitivity greater than 6°C per 4 W/m^2 is also extremely unlikely.[38] That said, uncertainty about the strength of climate feedbacks precludes dismissing the possibility that Earth will get surprisingly warm – climate feedbacks could turn out to be stronger than currently estimated.

Throughout this book, to compare various climate controls, we will use a central, probable estimate of climate sensitivity of 3°C per 4 W/m^2, or 0.75°C for every W/m^2.

2.3. Integrating systems, science, and policy

The Sun is the ultimate source of the energy in Earth's climate system. But what happens to that energy within the boundaries of our planet matters. If the energy coming from the Sun exceeds the energy leaving Earth, the temperature of the planet will rise until balance is restored. If more energy leaves than arrives, Earth will cool, again until energy balance is restored. Since human activity is perturbing energy flows, it stands to reason that human activities – such as regulations, adoption of energy-efficient technologies, and behavior change – could also bring balance to the system.

Recognizing that humans are an active component of the climate system, we attempt in this book to integrate climate science with policy responses that can address the climate change challenge effectively. For purposes of mitigation – slowing or preventing change at the source of perturbation – we can examine the different pathways by which energy travels in Earth's climate system, to see which ones we potentially could influence with deliberate action. For purposes of adaptation, understanding the flows of energy through Earth's climate system allows us to estimate likely future changes, so we can anticipate the conditions to which we will need to adapt. As you evaluate information here and elsewhere regarding human actions to respond to climate change, ask yourself whether the choices make sense from a systems perspective. Consider whether human actions might help amplifying or stabilizing feedbacks, or have consequences – possibly unintended – for parts of the system that we do not perturb directly. Consider whether we understand the system well enough to take a particular action.

CHAPTER THREE

Climate Controls: Energy from the Sun

3.1. Incoming solar radiation

- 3.1.1. Blackbody radiation: The Sun versus Earth
- 3.1.2. Our place in space: The Goldilocks planet

3.2. Natural variability

- 3.2.1. 4.5 billion years of solar energy
- 3.2.2. Orbital controls: Baseline variability in the past few million years
    - *3.2.2.1. Eccentricity: The shape of Earth's orbital path*
    - *3.2.2.2. Tilt*
    - *3.2.2.3. Precession of the equinoxes*
    - *3.2.2.4. The link to ice age cycles*
- 3.2.3. Sunspots: How important?

3.3. Response strategies

MAIN POINTS:

- The Sun gives off a great deal of shortwave radiation because it is hot; Earth gives off longwave radiation, and much less energy than the Sun, because it is cool.
- Solar energy output and the distribution of energy received over Earth's surface vary on long to short time scales.
- Geologically recent natural climate cycles have alternated between ice ages and warm periods. Human actions are now perturbing this background climate.
- The Sun is not responsible for global heating in the past fifty years.
- Response strategies to change the amount of solar energy reaching Earth through space are limited.

3.1. Incoming solar radiation

If there were a climate superhero, commanding the Sun would be her primary power. In Earth's climate system, the only source of energy of importance is energy from the Sun; there is a little bit of geothermal energy from Earth's interior, but it is negligible compared with solar energy. Most energy sources humans have harnessed through history ultimately originated from the Sun. Solar cells capture the Sun's energy directly and convert it to usable electricity, and wind – which spins turbines to generate electricity – is driven by differences in solar heating across the globe. Fossil fuels, too, ultimately are solar energy stored by plants long ago. Hydroelectric power is possible because solar energy evaporates water from the oceans, and Sun-driven winds blow clouds over land, where the rainwater can be captured or diverted. Even once commonly used whale oil would not exist without the marine food web, which exists because of photosynthetic organisms using the Sun's energy. The same is true for any biologically derived fuel, such as wood, dung, and charcoal.

In this chapter, we examine the role of the Sun, our primary energy source, in Earth's climate system. How much energy does it provide? Why is Earth, of all the planets in the solar system, ideally situated to support life? How has solar energy changed since Earth's formation? What was the Sun's role in the ice age cycles of the past few million years – the "baseline" climate regime in which humans evolved and against which we can compare the rapid climate changes we observe today? Can solar variability explain twentieth-century warming? (No.) Of the three main climate controls – solar energy, Earth's reflectivity, and the greenhouse effect – energy from the Sun is the one least likely to be changed by human actions. Later in the chapter, we also introduce the concept of geoengineering as a possible response strategy.

3.1.1. Blackbody radiation: The Sun versus Earth

All atoms at temperatures above absolute zero (0 kelvin, K) wiggle and vibrate and radiate energy. The chair you might be sitting on is radiating, the ice cream you might be eating, this book, and even you are radiating. The chair, book, ice cream, and you are also absorbing energy – we are bombarded with it all the time, from all directions. Consider each of these four items a little system, with energy outflow (the radiation emitted) and energy inflow (the radiation absorbed). If inflow and outflow balance, your chair, book, ice cream, and body will maintain their respective temperatures. Imbalances between energy in and energy out produce temperature changes.

Physicists have developed the concept of an ideal object, a "blackbody," that absorbs all the radiation that hits it and that emits an equal amount of radiation. In reality, chairs, books, ice cream, people, and stars and planets are not perfect blackbodies, but they are close. You do not absorb all the radiation that hits you – some of it gets reflected off your surface. If the radiation is in the form of X-rays, some of it might pass right through you. But the conceptual model of a blackbody helps us to grasp some basic important characteristics about the energy emitted by the Sun and the energy both absorbed and emitted by Earth. These characteristics are part of our climate reality. For the following discussion, we will assume that the Sun and Earth act like perfect blackbodies.

The two aspects of blackbody radiation that matter for climate are the kind of energy that is radiated, and the amount of energy that is radiated. By "kind" of energy, we mean wavelengths of energy. Figure 3.1 is a drawing of the electromagnetic spectrum. At one end of the spectrum is energy with very short

Figure 3.1. The Electromagnetic Spectrum

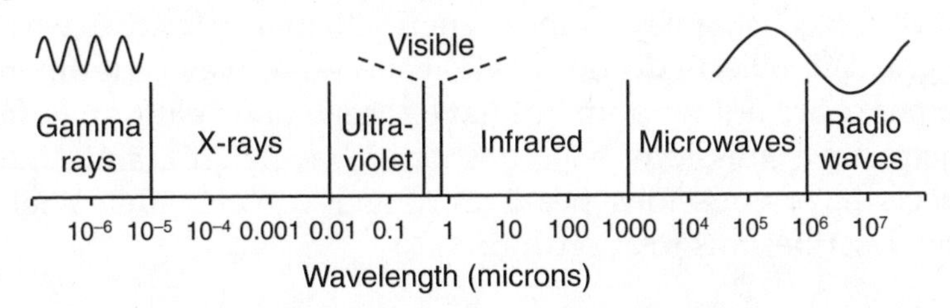

The ultraviolet, visible, and infrared parts of the spectrum are the most important for climate.

wavelengths, very high frequencies, and high energies per unit of radiation; examples include gamma rays, X-rays, and ultraviolet light. At slightly longer wavelengths is visible light. At wavelengths longer than visible light, there is infrared (IR), and then on to microwaves and radio waves. Objects at different temperatures radiate different amounts of energy at each of the wavelengths along the spectrum.

As for the "amount" of energy, hot objects radiate more energy than cold objects. You radiate more energy (per unit surface area) than does your ice cream, unless it has been heated to body temperature. Hot beach sand radiates more energy per square meter than does the nearby cool ocean water. For climate, it matters how much energy Earth absorbs, and therefore how much it must emit to balance energy outflow and energy inflow.

Insightful physicists – notably Max Planck and Wilhelm Wien – determined highly useful relationships between the temperature of a blackbody and both the kind and the amount of radiation the object emits. These relationships are handy for measuring the temperatures of objects such as the Sun. Although we cannot simply go stick a thermometer into the Sun, we can measure the kind and amount of energy it gives off, and then use that information to calculate its temperature.

Figure 3.2 shows how four idealized, blackbody objects at different temperatures radiate energy at different wavelengths – the "radiation spectrum" of each object. For example, the object at 320 K emits 27 W/m^2 (watts per meter squared) at a wavelength of 6 microns, while the object at 280 K emits only 9.3 W/m^2, and the object at 200 K emits hardly anything at this wavelength. Notice the peaks of each of the four curves: the wavelength at which the object gives off more energy than at any other wavelength. The objects do not peak at the same wavelength, and the peaks are not random but have a systematic pattern: hotter objects have emissions peaks at shorter wavelengths and colder objects have emissions peaks at longer wavelengths. Wien's Law describes this relationship:

$$\lambda_{peak} = 2898/T \tag{1}$$

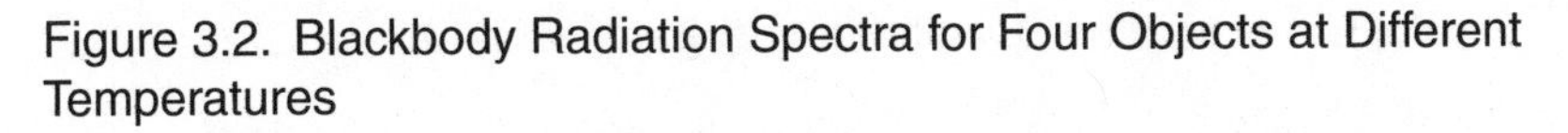

Figure 3.2. Blackbody Radiation Spectra for Four Objects at Different Temperatures

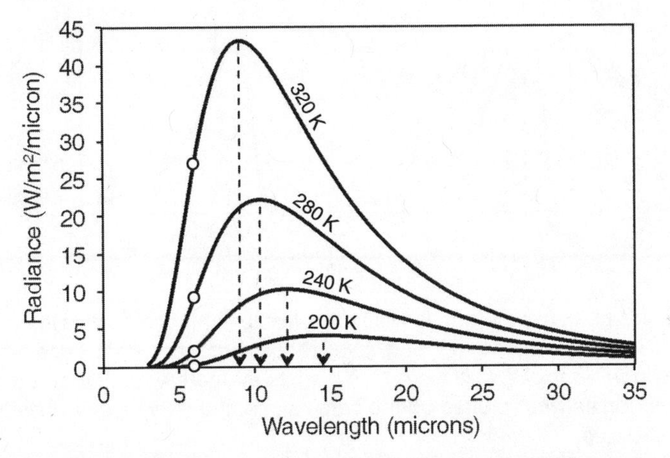

Notice that (1) at a particular wavelength – say, 6 microns – the hotter objects give off more energy (compare open circles); (2) the peak emission for hotter objects occurs at shorter wavelengths (tips of vertical arrows); and (3) the hotter the object, the more total energy it emits (imagine the area under each curve).

where λ_{peak} is the peak wavelength of a blackbody in microns (μm), T is the temperature of the blackbody in K, and 2898 is a constant with units of μmK. Notice that T and λ_{peak} are inversely related: as T goes up, λ_{peak} goes down, and vice versa. Look at Figure 3.2 to check.

So, we can measure the radiation (energy) emitted by the Sun, find its peak wavelength (λ_{peak}), then use equation (1) to figure out the Sun's surface temperature. The Sun's peak wavelength is at about 0.5 μm, which is in the middle of the visible light range; therefore – using equation (1) – the Sun's surface temperature is about 5,800 K. Try it backward. As we will see in detail later, Earth's average surface temperature is about 288 K (15°C), so λ_{peak} for Earth's surface is about 10 μm, which is in the infrared range. Earth is a lot colder than the Sun and emits radiation at longer wavelengths (Figure 3.3). What about λ_{peak}

Figure 3.3. Radiation Spectra for the Sun and Earth

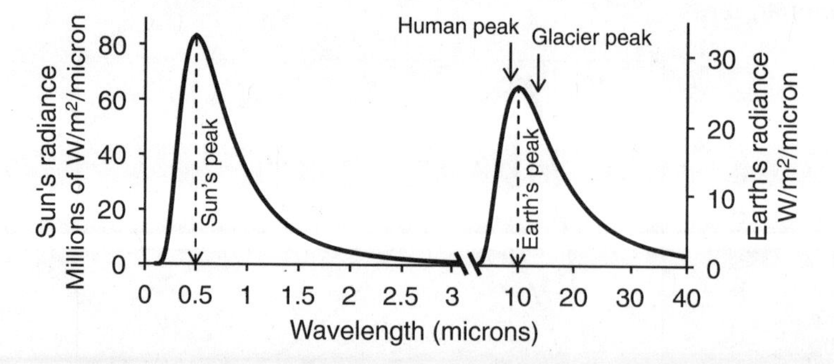

The Sun's peak wavelength is within the visible light, or "shortwave," range; Earth's is in the infrared, or "longwave," range. Note the vertical scale for the Sun is in *millions* of W/m^2/micron, as it emits far more energy than Earth; if the two bodies were plotted on the same scale, the curve for Earth would not even show up.

for an elephant, or for a glacier? How hot are humans? Pretty hot, but not hot enough to emit visible light.

Wien's Law leads us to a key point: energy given off by the hot Sun is dominantly "shortwave" and energy given off by the cooler Earth is dominantly "longwave." This difference in the types of energy given off by the Sun and Earth is one of the most fundamentally useful pieces of information to understand Earth's greenhouse effect. The reason certain gases are categorized as "greenhouse gases" is that they interact with particular wavelengths of radiation, some of which happen to match some of the energy emitted by Earth – that is, longwave radiation. The greenhouse effect is all about longwave radiation, *not* about shortwave radiation coming from the Sun, as we will see in Chapter 5.

Even though shortwave solar energy is not involved in the greenhouse effect, the Sun is our ultimate source of energy. Its output, combined with the orbital distance of Earth from the Sun, determines how much energy we get. How much? Handily, the total energy output of an object also depends on its temperature. In the 1800s, physicists – notably Jozef Stefan and Ludwig Boltzmann – determined the relationship between a blackbody's temperature and the total amount of energy it radiates, which we can express as:

$$F = \sigma T^4 \quad (2)$$

In this equation (the Stefan-Boltzmann Law), F is the energy radiated in W/m^2, σ is the Stefan-Boltzmann constant, named after those diligent physicists – it has a value of 5.67×10^{-8} $W/m^2 K^4$, and was figured out after many careful measurements – and T is, again, the object's temperature in K. Notice that temperature is raised to the power of four, which means that the energy emitted goes up exponentially as temperature goes up. If you compare two objects, one at 100 K and the other at 200 K, the warmer one gives off sixteen times more energy than the colder one, even though it is only twice as hot. Visually, in Figures 3.2 and 3.3, the total amount of energy, per square meter, given off by each object is the area under each curve. The hotter the object, the bigger the area under the curve and the more total energy given off.

The value of F tells us how much energy is given off per square meter of an object. Imagine all those one-meter-square picnic blankets from Chapter 2 laid out on the surface of the Sun (before they burn up). The area covered by each picnic blanket would give off σT_{Sun}^4, which turns out to be about 63,500,000 W/m^2. If we add up all the picnic blankets covering the Sun, the Sun's total output turns out to be about 3.9×10^{26} watts.

How much of that do we get here on Earth? That enormous quantity of energy given off by the Sun gets spread out over a larger and larger area as it travels outward – like the color in a balloon stretches out and fades as you blow it up. At the distance of Mercury's orbit, the Sun's 63,500,000 W/m^2 has been diluted to about 9,280 W/m^2. At Venus, it is down to about 2,650 W/m^2. And by the time the solar energy reaches the distance of Earth's orbit, we get about 1,365 W/m^2. This number, for Earth, is called the "solar constant," even though, as we will see in section 3.2., it is not quite constant.

If you look back at Figure 2.6, however, you will see the incoming solar radiation is not 1,365 W/m^2, but 341 W/m^2. This is because the Earth is spherical and it spins, so the incoming solar radiation is spread over the entire planet's surface area. Of course, the tropics get more than the poles, the summer hemisphere gets more than the winter hemisphere for a few months, and only the "day" side gets radiation at any particular time, so the 341 W/m^2 is the annual average for the whole planet.[1]

If the Earth gets an inflow of 341 W/m^2 of solar radiation, it has to have an outflow of the same amount if it is to be in energy balance. Look again at Figure 2.6. Notice that, on the left-hand side, some of the incoming solar radiation (102 W/m^2, or about 30 percent of the total) reflects off material in the atmosphere and off Earth's surface and goes right back into outer space. So, after accounting for this reflection outflow, we are left with 239 W/m^2 that has to leave Earth's system, and since space is a vacuum, radiation is really the only available mechanism to transfer that energy out.

How hot does Earth have to be to radiate away 239 W/m^2? To determine this, we can use equation (2) in reverse: $T_{Earth} = (F/\sigma)^{0.25} = [(239\,W/m^2/5.67 \times 10^{-8}\,W/m^2K^4)]^{0.25} = 255\,K$, or −18°C, which is Earth's "effective radiating temperature." Fortunately, that is not Earth's average surface temperature, but the average temperature that the upper atmosphere, where energy is radiating directly back into space, has to be for the Earth to remain in energy balance. Naturally, there are some complications. Some of the energy that makes it back into space comes from lower in the atmosphere; some comes from Earth's surface, but the −18°C helps us grasp the magnitude of warming that the greenhouse effect provides. As we noted earlier, Earth's average surface temperature is about 15°C. Without greenhouse gases, it would be about −18°C, so, at this point in Earth's history, the greenhouse effect provides about 33°C of warming.

Before leaving this concept, let us use equation (2) one more time. How much energy does Earth's surface emit if it averages 15°C? The answer is $F_{surface} = \sigma T_{surface}^{\ 4}$ = about 390 W/m^2, which is more than the solar radiation absorbed by Earth's surface. Why does Earth's surface have to radiate more energy than the solar energy it absorbs? The reason is that Earth's surface has another inflow. Greenhouse gases in the atmosphere absorb some of the longwave radiation coming from Earth's surface. They re-emit this energy in all directions, some of it back downward towards Earth's surface, some upward towards space. Earth's surface absorbs some of the downward-heading energy, which adds to the solar energy that Earth's surface has to get rid of to be in energy balance. Total outflows must match total inflows, both at Earth's surface and for the Earth system as a whole. If they do not, Earth's temperature will change.

In summary, because the Sun is hot, it emits lots of energy per square meter. Most of that energy is relatively short wavelength radiation on the electromagnetic spectrum. Earth, which is cooler, emits less energy per square meter, and most of that energy is at longer wavelengths. The Sun, with its peak wavelength in the visible light range, shines brightly. Earth, with its peak wavelength in the infrared, gives off energy we cannot see.

3.1.2. Our place in space: The Goldilocks planet

If, someday, humans leave Earth to find a new habitable planet, we might find ourselves in the position of Goldilocks in the story "Goldilocks and the Three Bears." Goldilocks tested porridge until she found a bowl with a temperature that was "just right." Earth, too, has features that make it "just right" or, at least, within an acceptable range, for habitation. This is not entirely surprising given that life as we know it evolved in the conditions that exist on Earth. But because it is "just right," Earth is sometimes called the "Goldilocks planet," though it might be more accurately named the "littlest bear's porridge planet."

Compare Earth to Venus and Mars, the rocky planets nearest us, both of which are inhospitable to life. As we have seen, Earth's effective radiating temperature is –18°C, but with the greenhouse effect, the surface temperature is about 15°C. Venus, closer to the Sun but with a much higher reflectivity, has an effective radiating temperature of –43°C, but an overwhelming greenhouse effect. Virtually all the carbon on Venus is in its atmosphere as CO_2, leading to a surface temperature of about 460°C. Mars, farther from the Sun, receives less solar energy and, like Venus, its atmosphere is mostly CO_2. But the Martian atmosphere is so thin that its greenhouse effect is very small, so the planet's average surface temperature is just a few degrees above its effective radiating temperature of –62°C.

The temperature range on Earth is such that water – a crucial ingredient for life – exists in liquid, gaseous, and solid forms. Venus, however, got too hot long ago, and all its water escaped into space. Mars has water, but apparently not very much, and the little it has seems to be frozen most of the time. Earth's interior is hot enough to provide energy to drive plate tectonics on the surface, which provides long-term stabilizing controls on the greenhouse effect. Earth also has a magnetic field, which provides extra protection from the solar wind. And Earth's orbit is only a tiny bit eccentric, the planet rotates once every twenty-four hours, so most parts of it get frequent Sun exposure, and its 23.5° tilt produces reasonable seasonal extremes. Venus is about the same size as Earth and also has a fairly circular orbit and a small tilt, but it lacks plate tectonics and a magnetic field, and rotates only once every 225 Earth days. Mars is smaller than Earth and, like Venus, lacks plate tectonics and a magnetic field.

In short, if Goldilocks were stumbling around the solar system checking out planets for room and board, she would find Venus too hot, Mars too cold, but Earth just right.

3.2. Natural variability

In this book, we are concerned with climate change and mitigation and adaptation strategies on human time scales – tiny instants of geologic time. To align these strategies with climate reality, we need information about what drives climate. Currently, humans are conducting an experiment by putting lots of carbon into the atmosphere within a short period. Are there examples from the past, before humans evolved, in which something remotely similar happened? What natural factors drove the change? How did the climate system respond? Much effort today goes to forecasting the probabilities of various future climate change scenarios. Those forecasts come from models constructed to simulate how Earth works, including natural changes in solar energy. If the models decently represent Earth's climate system (see Chapter 7), they can help us understand which system parts bear responsibility for driving climate.

What climate changes can we attribute to the Sun? This section addresses how Earth's total solar energy inflow, and where the energy goes, varies on time scales from billions of years to seasons, to help us situate the Sun within the climate system.

3.2.1. 4.5 billion years of solar energy

Stars form, live, and then die. All the stars out there in the universe are at particular stages of star life. Based on observations and the modeling of star behavior, our star, the Sun, is middle aged – about halfway through its expected lifetime. Since the Sun's formation about 4.5 billion years ago, it has been converting hydrogen to helium through nuclear fusion and releasing energy, a tiny portion of which is intercepted by Earth. Over time, as more hydrogen fuses to helium, the Sun's core contracts, its temperature rises – which causes its radius to expand, compensating for core contraction – and it shines brighter. For our purposes, the important consequence of these gradual changes is that the Sun's energy output (W/m^2) is increasing slowly over time (see Figure 3.4). Three and a half billion years ago, as a young star, the Sun radiated less than 80 percent of the energy it radiates today. By about 70 million years ago, when dinosaurs were living, the Sun's energy output was about 99 percent of today's value.

If this long-term change in solar energy were the only control on Earth's climate, Earth's temperature would have gradually increased over its life. Instead, Earth's surface temperature has gone both up and down over time (Figure 3.5). Around 300 million years ago, Earth was pretty cold, but by 200 million years ago, it was ice free and warm. Recently, during the time when humans evolved, it has been cold again. The Sun alone could not have done this; clearly, the other two climate controls – reflectivity (Chapter 4) and the greenhouse effect (Chapter 5) – must play

Figure 3.4. The Sun's Energy Output Over Time

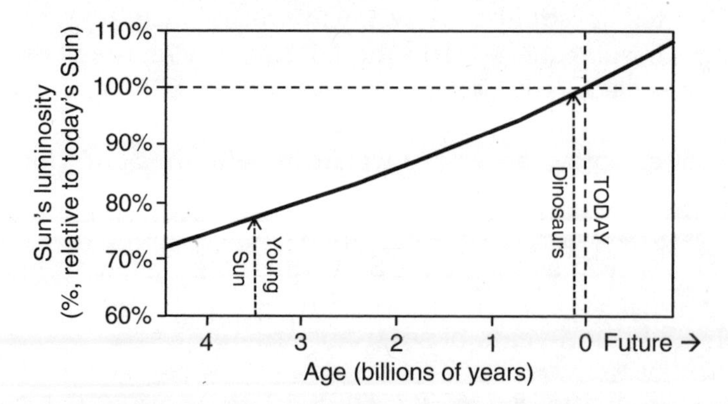

Source: Modified from I. Ribas, "The Sun and Stars as the Primary Energy Input in Planetary Atmospheres," *Proceedings of the International Astronomical Union* 5 (2009): 3–18, figure 1. Reproduced with permission of Cambridge University Press.

major roles. The Sun's changes over billions of years brought us to today's solar constant of 1,365 W/m^2. We expect this long-term average value to increase over time, but so slowly that it will not drive climate on human time scales.

3.2.2. Orbital controls: Baseline variability in the past few million years

Currently, two places on land are covered with great quantities of ice: Greenland and Antarctica. Twenty thousand years ago, during the most recent ice age, ice sheets thousands of meters thick also covered most of the northern half of North America, plus parts of northern Asia and Europe. The global average temperature was 5–6°C cooler than today, and sea level was about 120 meters lower. Locations that are now on the coast would have been far inland. Our ancestors and the animals they hunted – such as wooly mammoths – were dealing with a different set of climatic circumstances.

Figure 3.5. Estimates of Earth's Temperature and Atmospheric CO_2 Concentrations over the Past ~600 Million Years

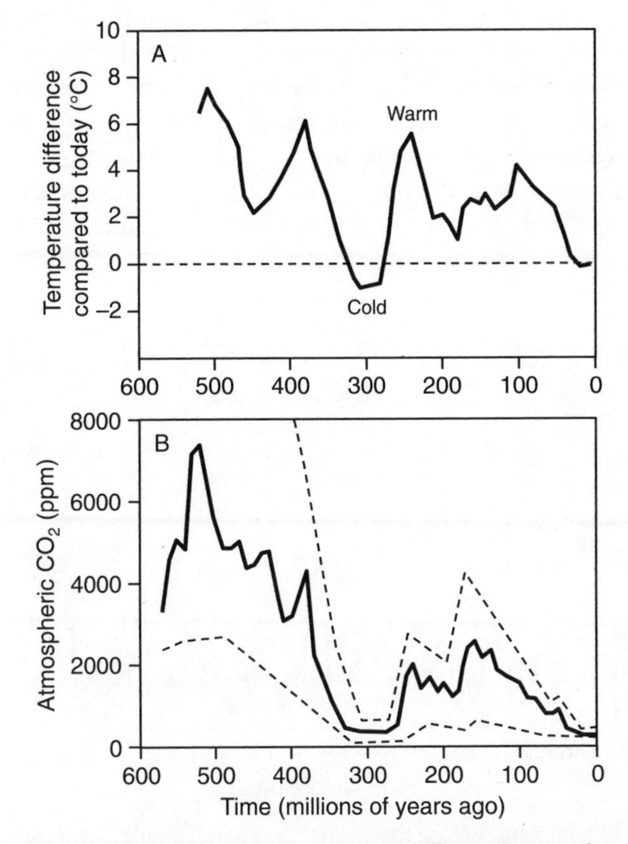

Panel A shows estimates of global temperature differences over time compared with today (dashed line at zero represents "today"); panel B shows estimates of atmospheric CO_2 within a probable range (dashed lines). The temperature data are better correlated with greenhouse gas concentrations than with the increasing solar energy output shown in Figure 3.4.

Source: Modified from D.L. Royer, R.A. Berner, I.P. Montañez, N.J. Tabor, and D.J. Beerling, "CO_2 as a Primary Driver of Phanerozoic Climate," *GSA Today* 14, no. 3 (2004): 4–10, figures 2 and 4 .

The most recent ten thousand years have been a warm period in a series of cold-warm, glacial-interglacial cycles extending back a few million years (Figure 3.6 shows the most recent 800,000 years). We can consider these cycles the "natural background" in which we exist today and against which we can compare present-day climate changes. Why do these climate cycles happen?

Gravitational attractions among the Earth, the Sun, and major planets cause small, predictable variations in Earth's orbital geometry. The three parameters of interest here are

Figure 3.6. Estimates of Earth's Temperature and Atmospheric CO_2 Concentrations over the Past 800,000 Years

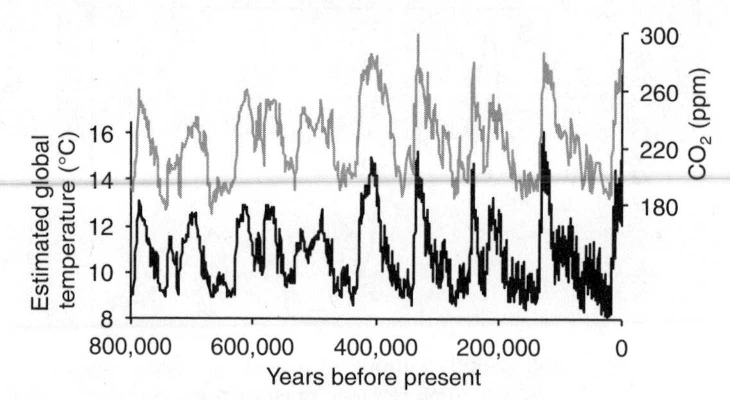

Records of past temperature (black line) and atmospheric CO_2 (gray line) are preserved in ice sheets. These data were obtained from Antarctic ice cores, which contain samples extending back about 800,000 years. Both temperature and CO_2 show distinctive cycles over time.

Sources: Temperature data from J. Jouzel, V. Masson-Delmotte, O. Cattani, G. Dreyfus, S. Falourd, G. Hoffmann, B. Minster, et al., "Orbital and Millennial Antarctic Climate Variability over the Past 800,000 Years," *Science* 317 (2007): 793–796, scaled as in J. Hansen, M. Sato, P. Kharecha, D. Beerling, R. Berner, V. Masson-Delmotte, M. Pagani, et al., "Target Atmospheric CO_2: Where Should Humanity Aim?" *Open Atmospheric Science Journal* 2 (2008): 217–31. Composite CO_2 data from D. Lüthi, M. Le Floch, B. Bereiter, T. Blunier, J.-M. Barnola, U. Siegenthaler, D. Raynaud, et al., "High-resolution Carbon Dioxide Concentration Record 650,000–800,000 Years before Present," *Nature* 453 (2008): 379–82.

(1) the shape of Earth's orbit around the Sun, (2) the tilt of Earth's axis, and (3) the position along Earth's orbital path at which different seasons occur. Variations in these three parameters are cyclic, and can be calculated into both the past and the future. These orbital changes are also called "Milankovitch cycles," after the Serbian astronomer, Milutin Milankovitch, who first computed their variability over time.

It is not surprising that the circumstances of Earth's geometry and orbit around the Sun influence climate. We know, for example, that Earth's tilt causes the annual seasons[2] and that higher latitudes experience more marked contrasts between summer and winter than low latitudes. Slow changes in Earth's orbital geometry alter the total amount (slightly) and the distribution (more important) of solar radiation the planet receives over time. Notably, orbital variability alters the summer-winter seasonal contrast at particular latitudes by changing the distribution of energy over the planet's surface. High seasonal contrast means extra warm summers coupled with extra cold winters; low seasonal contrast means mild, cooler summers and mild, warmer winters. Northern high latitudes, where ice sheets form, matter most. Take a guess now at which seasonal contrast scenario would help an ice sheet to grow or melt. After we see how Earth's orbit affects seasonal contrast, we will return to this question. In short, orbital changes – assisted by processes on Earth – have driven the glacial-interglacial cycles of the past few million years.

3.2.2.1. Eccentricity: The shape of Earth's orbital path

Every year, Earth does one lap around the Sun. If its path were perfectly circular, the distance between Earth and the Sun would always be the same. Instead, the path is slightly elliptical, and the Sun is at one focus of the ellipse, such that, as Figure 3.7, panel A shows, at some times during the year Earth is slightly farther from the Sun (currently, the farthest point is in July) and at other times it is slightly closer (currently, the closest point is in January). We saw in section 3.1. that one of the

Figure 3.7. Earth's Orbital Parameters

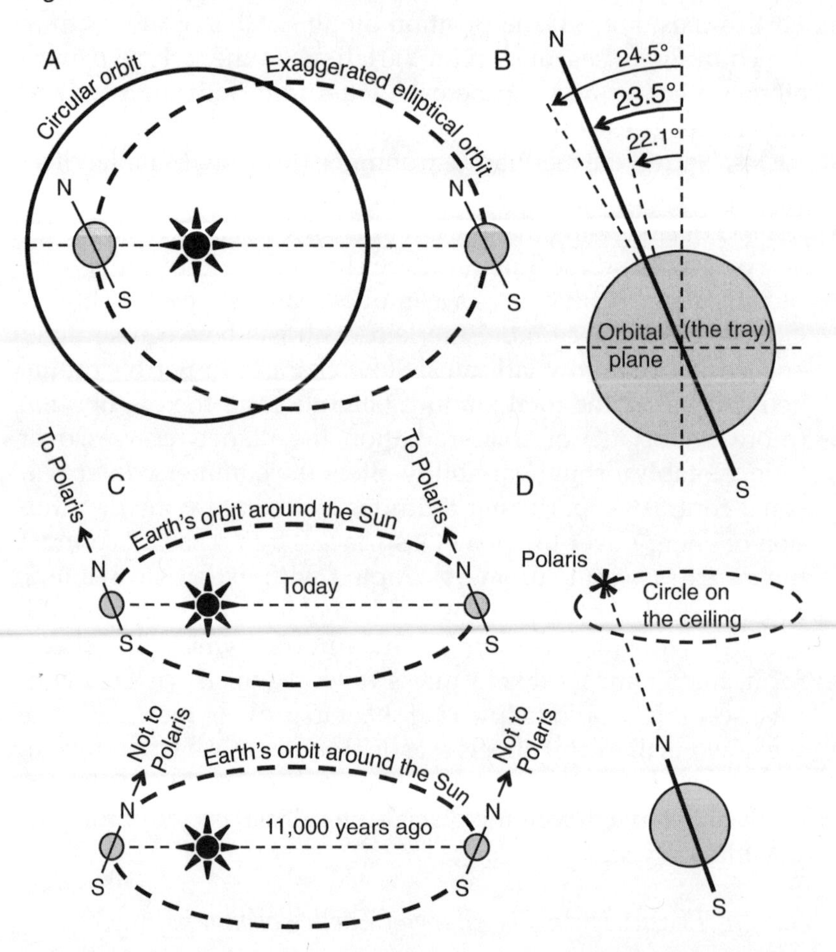

Panel A shows that the shape of Earth's orbit around the Sun varies from nearly circular (solid line) to slightly elliptical (thick dashed line); the drawing is exaggerated. Panel B shows that Earth is tilted over about 23.5° from vertical, and the tilt varies between 24.5° and 22.1°. Panel C shows the precession of the equinoxes, with today's orbital configuration and that of 11,000 years ago. Panel D shows how Earth's axis points to different places in space over time, causing the precession of the equinoxes to occur.

important parameters determining how much solar radiation Earth receives is its distance from the Sun. Through the course of a year, this amount varies a little bit, because the Earth-Sun distance varies a little bit. But, more important, the shape of the orbital path changes over time, too: the circle gets a little more squashed, then a little less squashed, back and forth – in geometry terms, the "eccentricity" of Earth's orbit varies. The time it takes to complete a full cycle from maximum (most squashed) to minimum (most circular) eccentricity and back to maximum is about 100,000 years.

The annual average solar radiation Earth receives when the orbit is more eccentric is about 0.4 W/m^2 higher than when it is less eccentric. That is not much difference – think of it as potential variability around the 341 W/m^2 in Figure 2.6. Of the three orbital parameters, eccentricity is the only one that alters the absolute amount of solar energy Earth receives over time, but the change is small and slow. The other two, however – Earth's tilt and the precession of the equinoxes – change the spatial distribution of that energy.

3.2.2.2. Tilt

Imagine the ellipse describing Earth's orbit around the Sun as the edge of a tray. If Earth had no tilt, its rotational axis – the north-south axis around which the planet spins – would stand upright relative to the tray's surface. Instead, the axis is tilted over about 23.5° (see Figure 3.7, panel B). The north end of Earth's rotational axis today is pointed out into space towards Polaris, the North Star. It keeps pointing towards Polaris all year, regardless of what side of the Sun we happen to be on. During times of the year when Polaris is on the opposite side of the Sun from us, the northern hemisphere is tilted towards the Sun and the southern hemisphere is tilted away. The north has summer and the south has winter at that time. Six months later, when we are back on the Polaris side of the Sun, the northern hemisphere is tilted away from the Sun and has winter, while the southern hemisphere is tilted towards the Sun and has summer.[3]

In short, Earth's tilt is the reason we have seasons, and explains why the seasons are opposite in the northern and southern hemispheres. Like its eccentricity, Earth's tilt varies a little bit over time: sometimes the planet is tilted over a little bit more, as much as 24.5°, and sometimes it is a little more upright, with a tilt as small as a 22.1° (Figure 3.7, panel B). The tilt varies smoothly, taking about 41,000 years to move through the full cycle from maximum tilt to minimum tilt and back. A planet with no tilt would have no seasons. Venus, for example, is nearly upright and has very little seasonal contrast; Uranus, at the other extreme, has a nearly horizontal tilt, so that, during its northern hemisphere's summer, practically the entire southern hemisphere is in the dark, giving the planet a huge seasonal contrast. Earth's 23.5° tilt today is currently decreasing (getting more upright), meaning that, according to tilt alone, the planet is headed towards lower seasonal contrast.

3.2.2.3. Precession of the equinoxes

But tilt is not the only control on seasonal contrast. Given Earth's slightly elliptical orbit, we are slightly closer to the Sun during the southern hemisphere's summer (northern hemisphere's winter) and slightly farther away during the southern hemisphere's winter (northern hemisphere's summer). This combination means the southern hemisphere has higher seasonal contrast today – in terms of incoming solar energy – than does the northern hemisphere (Figure 3.7, panel C). About 11,000 years ago, however, Earth passed the closest point to the Sun not in January, as it does today, but in July, during the northern hemisphere's summer. And all the other months get their turn. This effect is called the "precession of the equinoxes." The equinoxes are the two times of the year when the two hemispheres get equal lengths of day and equal solar exposure. The precession of the equinoxes describes the pattern by which these points shift along the orbital path over time. Occasionally, the equinoxes happen to coincide with the closest and farthest points on Earth's orbit, but most of the time, they are somewhere in between these special points. The precession of the equinoxes could just as well have been called the precession of the solstices, or the precession of November, but the equinoxes are a good choice – those special times when the hemispheres are equal.

How does it work? The most common analogy is a spinning top. If you spin a toy top, notice that it spins quite quickly – like the daily spin cycle of Earth – and its axis does not always point in the same direction. If you were to attach a long pencil to the axis of the top, and if the pencil could draw on the ceiling as the top spins, it would draw some semblance of a circle (Figure 3.7, panel D). Then, pick a point on that imaginary circle and label it "Polaris," to indicate where Earth's north polar axis points today. The fact that the axis points towards Polaris today is just a coincidence, and incredibly convenient for navigators in the northern hemisphere. At times in the past, and times in the future, the North Pole has pointed, and will point, towards all the other points along the circle: sometimes it points to a star (both Thuban and Vega have been the "North Star" in the past), but most of the time, it points towards a dark part of the sky. For the pencil to draw the full circle on the ceiling – for the equinoxes to march all the way around Earth's orbital path – takes about 21,000 years (actually, a somewhat complex combination of cycles operating at about 19,000 and 23,000 years).

Notice that, if Earth's orbital path were perfectly circular – that is, if its eccentricity = 0 – the precession of the equinoxes would not change the seasonal contrast between the hemispheres at all. The effect of precession is largest when the orbital path is most eccentric, and virtually nothing when the orbit is close to circular.

3.2.2.4. The link to ice age cycles

How does all this orbital geometry relate to Earth's climate? Changing the tilt changes the seasonal contrast in both hemispheres at the same time. In contrast, precession, modulated by eccentricity, makes the two hemispheres have opposite extremes in seasonal contrast at the same time. The actual seasonal contrast for different parts of Earth turns out to be a somewhat messy combination of eccentricity, tilt, and precession, all operating at different frequencies. In terms of W/m^2, the average energy received at some particular latitude in some particular season – say, 60°N in July – could vary by about 30 W/m^2 over 10,000 years. Compare this with the mere 0.4 W/m^2 of maximum variability in total annual solar energy Earth receives. The comparison highlights that these orbital parameters have only a small effect on the total energy Earth receives, but a large effect on how that energy gets distributed over Earth's surface.

What would be the best configuration to help grow an ice sheet in the northern hemisphere – high or low seasonal contrast? Consider high seasonal contrast first: exceptionally hot summers and cold winters. Perhaps cold winters would produce lots of snowpack, but hot summers could melt it all. What about low seasonal contrast? Mild winters could still be cold enough to have snowfall, but the following cool summer might not be warm enough to melt away all the previous winter's snowfall, leaving a base to build on the following winter. Once the snow cover got going, the white snow and ice would reflect some of the Sun's incoming energy, helping to preserve the existing ice, decreasing the amount of solar energy Earth absorbs, cooling Earth's temperature, and starting an amplifying feedback that promotes continued ice sheet growth. This scenario, with low seasonal contrast in the northern hemisphere promoting ice sheet growth, turns out to match Earth's climate history.

The amplifying feedback works in reverse, too. As ice starts to melt, it exposes more dark land, which absorbs more solar radiation, promoting warming, promoting ice melt, and so on. Other amplifying feedbacks on Earth are also important. For example, a little warming can trigger a little outgassing of CO_2 from the oceans, adding to the greenhouse effect, promoting warming, promoting more outgassing. It turns out that the orbital changes are key, but their role is mostly as timekeeper, telling the players when a new game starts and when it is half-time. Responses and feedbacks within Earth's climate system, nudged by orbital changes in solar radiation, are what produced the large amplitude glacial-interglacial cycles of the past few million years. Barring some catastrophe, such as a massive collision of Earth with an extraplanetary body, the orbital cycles will march on as they have in the past. Earth's climate system will continue to respond, and feedbacks will continue to amplify and stabilize the response. These cycles are the natural background against which the climate system now operates.

Earth looked quite different during the last ice age, when it was a mere 5–6°C cooler. Today, global temperature is rising. Many governments have agreed that a rise of more than 2°C above pre-industrial temperatures would constitute "dangerous climate change" – "dangerous," that is, in terms of the implications for *humans* and for our development. At the time of this writing, we are already almost halfway to the 2°C increase mark, and a rise larger than 2°C appears well within the realm of possibility. Whether we exceed this threshold depends largely on future human actions. The planet, and life in one form or another, will be around for millions of years to come essentially no matter what we do. But the important point is that a small-sounding temperature rise of 5–6°C took the world from ice age conditions to the relatively pleasant and stable warmth we have enjoyed for the past 10,000 years. What would even half an equivalent change in further heating produce?

3.2.3. Sunspots: How important?

Due to nuclear reactions in the Sun's interior, its energy output has increased over billions of years. Due to changes in Earth's orbit over time, the total solar energy the planet receives changes by a little bit every 100,000 years. What about shorter time scales? Over years to centuries, the Sun's energy output varies depending on the number and size of sunspots – dark areas – on its surface. Typically, the number of sunspots cycles up and down about every eleven years. You might think that more dark spots on the Sun's surface would decrease its overall energy output, but the reverse is true. Although the dark center of a sunspot emits less than the average amount of energy per square meter, the area around each sunspot is brighter than average. The net result is more total energy output when there are more sunspots (Figure 3.8).

Since 1979, we have used instruments on satellites to measure sunspots and solar output. Before then, sunspot data come from telescopic observations, which allow us to reconstruct solar activity back to the early 1600s, when the telescope was invented.[4] For periods even farther in the past, we can use chemical records preserved in ice sheets, sediments, and tree rings – in particular, the amounts of beryllium-10 (^{10}Be) and carbon-14 (^{14}C) that were present over time. Through a series of relationships involving the solar wind, Earth's magnetic field, and cosmic rays, the accumulation of these isotopes varies inversely with the number of sunspots.[5]

How much do sunspots affect solar energy output? Recent sunspot cycles, for which we have very good measurements, have changed solar output, and energy received by Earth, by about 0.1 percent between the minimum and maximum of an individual cycle.[6] That would change Earth's 341 W/m^2 of incoming solar radiation by just 0.34 W/m^2 over the course of about eleven years. Longer term, since the 1600s, increased average sunspot activity has increased the average solar energy received also by about 0.1 percent.[7]

Sunspots are the only factor we have considered so far that change Earth's energy budget on time scales of decades to centuries. Can changes in sunspots account for Earth's observed temperature changes over the past century? As Figure 3.8 shows, for the first half of the twentieth century, temperature records correlate pretty well with incoming solar energy: increasing sunspots and increasing temperature happened with similar timing. But since about 1960, the two records clearly have diverged: average global temperature has continued on an upward trajectory, while sunspot activity has flattened out and decreased slightly. If solar energy output were the primary culprit controlling Earth's average temperature, we would have observed cooling over the past fifty years. This fundamental mismatch between the temperature record and the sunspot record clears the Sun of responsibility for the recent warming trend.[8]

Figure 3.8. Sunspots, Incoming Solar Energy, and Changes in Earth's Temperature since 1880

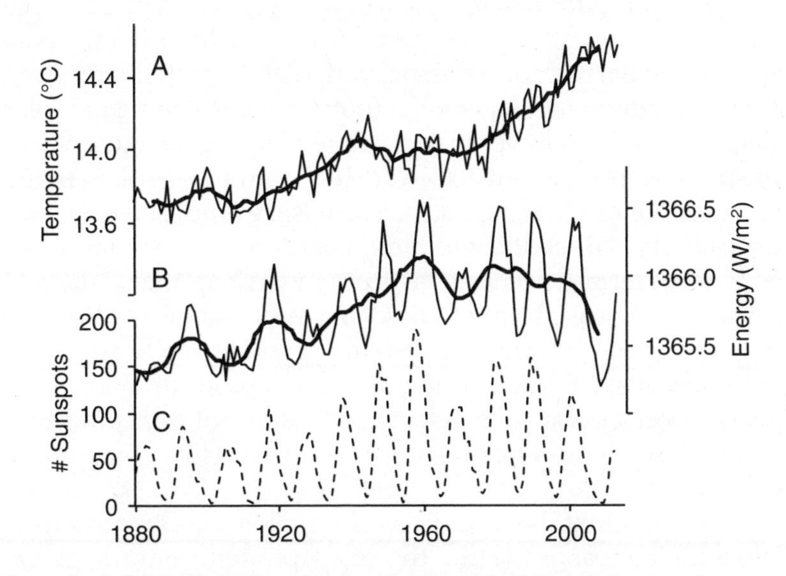

Changes in incoming solar energy cannot explain the global average temperature rise since the middle of the twentieth century. Panel A shows the global annual average surface temperature since 1880 (thin line), with an 11-year running average (thick line); panel B shows incoming solar energy (total solar irradiance), which is the variability in the "solar constant" (thin line), with an 11-year running average (thick line); panel C shows the number of sunspots over the period.

Sources: Panel A data from United States, National Aeronautics and Space Administration, Goddard Institute for Space Studies. Panel B data (for the period 1880–1978) from N.A. Krivova, L.E.A. Vieira, and S.K. Solanki, "Reconstruction of Solar Spectral Irradiance since the Maunder Minimum," *Journal of Geophysical Research* 115, no. A12 (2010); and (for 1979–2012) from Physikalisch-Meteorologisches Obervatorium Davos World Radiation Center, unpublished data from the VIRGO Experiment on the cooperative ESA/NASA Mission SoHO, version d41_62. Panel C data from United States, National Aeronautics and Space Administration, Marshall Space Flight Center, "Solar Physics," available online at http://solarscience.msfc.nasa.gov/greenwch/spot_num.txt.

This is not to say that sunspots are irrelevant. Sunspots do change the amount of energy Earth receives. The Little Ice Age in the 1600s to 1800s, when parts of the northern hemisphere were particularly cool, is associated with a particularly long stretch of time when sunspot activity was low, and thus solar energy output was lower than average.[9] During the most recent sunspot low, from about 2007 to 2010,[10] a little less solar energy than average reached the Earth. Likewise, the next peak in sunspot activity will contribute more energy to the system, lending a short-term helping hand to the warming that is largely driven by increased emissions of greenhouse gases. Because of delays in Earth's climate system – particularly heat uptake by the oceans – Earth's temperature cannot go up and down quickly in response to the eleven-year sunspot cycle, but longer-term trends in sunspots do influence Earth's climate and temperature.[11]

It turns out that the most powerful tool of the climate superhero is not responsible for the late-twentieth-century global heating. That responsibility, as we will see in detail later, lies largely with human emissions of greenhouse gases – particularly CO_2, which persists in the atmosphere for a long time. The Sun has a role in every climate scene, but in this recent case plays an antagonistic, rather than a supportive, character.

3.3. Response strategies

As we have seen, energy from the Sun is the key ingredient in the climate we all experience today, but tiny variations in its output – because of sunspot activity, for instance – are not the main causes of the climate change we observe. Creative and ambitious people around the world are now searching for solutions to climate change – strategies that might mitigate the problem or tackle it at its source. Since we have been dealing with solar input in this chapter, now is a good time to introduce a set of strategies that many people might slot into the "wacky" category: geoengineering. Wacky or not, geoengineering is getting increasing air time in the global conversation about climate change, in part because of spectacular uncertainties and ethical complications associated with it.

Geoengineering refers to intentional attempts to manipulate Earth's climate directly, and since it walks the traditional line between adaptation and mitigation, experts are beginning to think about it as a "third pillar" of responding to climate change. Geoengineering differs from traditional mitigation that focuses on greenhouse gases because it purportedly would allow us to keep pumping emissions into the atmosphere, but would intervene to "engineer" the climate. It also differs from adaptation because we would not be protecting ourselves from the effects of climate change, but tinkering with the system so that those effects do not occur (or at least are softened).

Some geoengineering strategies would influence the quantity of greenhouse gases in the atmosphere. An example of this is fertilizing the oceans so that algae can grow and multiply, thereby taking in more carbon dioxide from the atmosphere (for more information on these types of mitigation strategies, see Chapter 6). Other strategies would target Earth's energy budget, or its reflectivity. An example is injecting particles into the atmosphere that would help to block out part of the Sun's rays.[12] In theory, this could counteract the effect of a greater quantity of greenhouse gases in the atmosphere without actually reducing emission levels or concentrations (we deal with this option in more detail in Chapter 4).

You might think that there is not much we can do about changing the solar constant – the amount of energy that reaches Earth's upper atmosphere. But some scientists have proposed methods of "solar radiation management" that would enhance the reflectivity of the upper atmosphere, or even catch the radiation before it reaches us. One such proposal involves sending what is essentially a giant umbrella or sunshade – consisting perhaps of trillions of small discs or a single giant "lens"[13] – into orbit around 1.5 million km above Earth. The possible location for this sunshade would be at Lagrange point L1, a point between Earth and the Sun where their respective gravitational forces cancel one another and the object would be held in a relatively stable orbit.

Advocates suggest geoengineering is desirable for several reasons: (1) it would be cheap, at least compared with weaning the entire planet off fossil fuels; (2) it could "turn down" the planet's temperature quite quickly; and (3) it might be necessary, since we are not now doing a very good job of reducing emissions.[14] Trade-offs and complications from such a strategy abound, however. Who would control the global thermostat? What effect would tinkering with incoming solar radiation have on precipitation patterns, monsoons, wind, and other elements of the climate system? If we got it wrong, could we fix our mistake? Who should pay for it? We deal with some of these questions as we introduce geoengineering strategies throughout this book, and we touch on the ethical conundrums in Chapter 11.

Color Plates

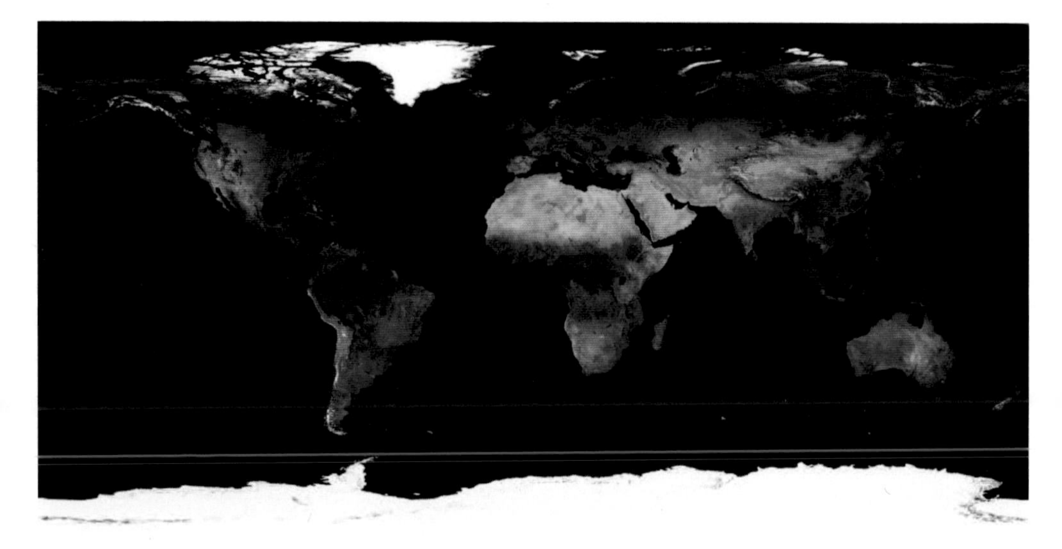

Plate 1. Composite satellite image showing what Earth's surface would look like without clouds. Different materials on Earth's surface have different albedo. Compare light-colored deserts and ice to dark-colored oceans and forests.

Source: United States, National Aeronautics and Space Administration, Earth Observatory (Greenbelt, MD); available online at http://eoimages.gsfc.nasa.gov/images/imagerecords/74000/74418/world.topo.200408.3x5400x2700.jpg. Data from August 2004.

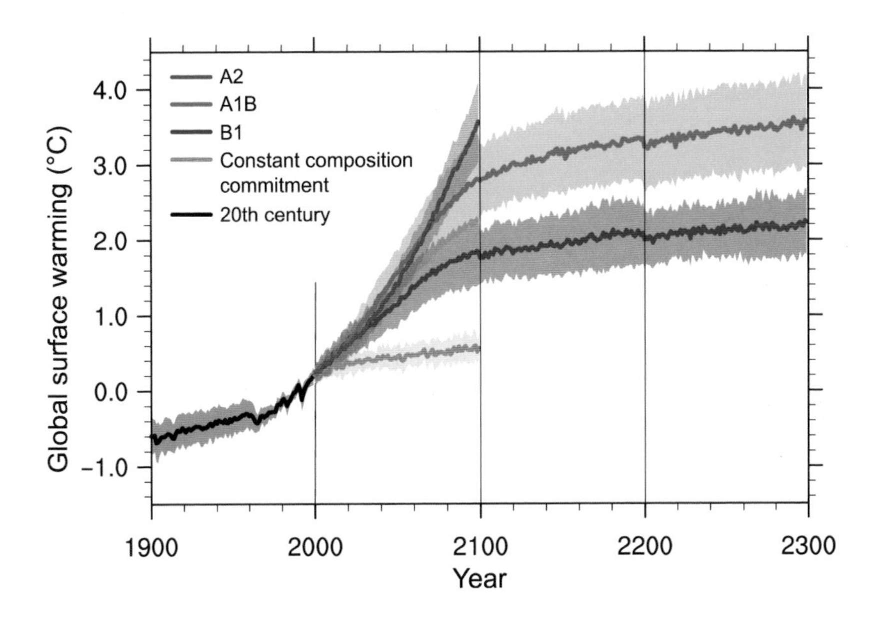

Plate 2. Mean temperature rise associated with the SRES emissions scenarios. Constant composition commitment refers to the amount of warming that will occur even if the composition of the atmosphere does not change further.

Source: Intergovernmental Panel on Climate Change, *Climate Change 2007: The Physical Science Basis. Contribution of Working Group I to the Fourth Assessment Report of the Intergovernmental Panel on Climate Change*, ed. S. Solomon, D. Qin, M. Manning, Z. Chen, M. Marquis, K.B. Avery, M. Tignor, and H.L. Miller (Cambridge: Cambridge University Press, 2007), figure 10.4.

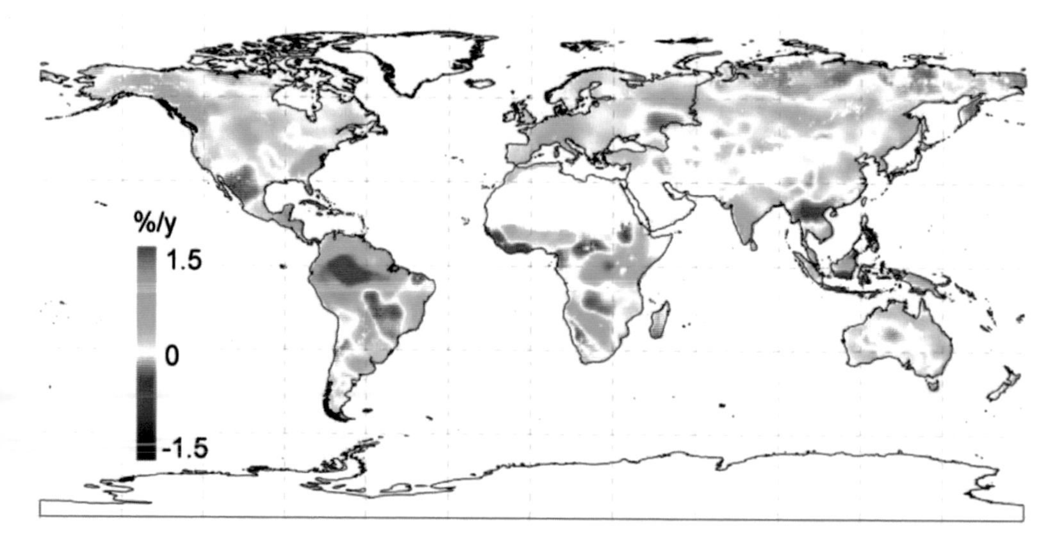

Plate 3. Estimated global change in terrestrial net primary productivity over the 1982–99 period.

Source: R.R. Nemani, C.D. Keeling, H. Hashimoto, W.M. Jolly, S.C. Piper, C.J. Tucker, R.B. Myneni, et al., "Climate-driven Increases in Global Terrestrial Net Primary Production from 1982 to 1999," *Science* 300, no. 5625 (2003): 1560–3. Reprinted with permission.

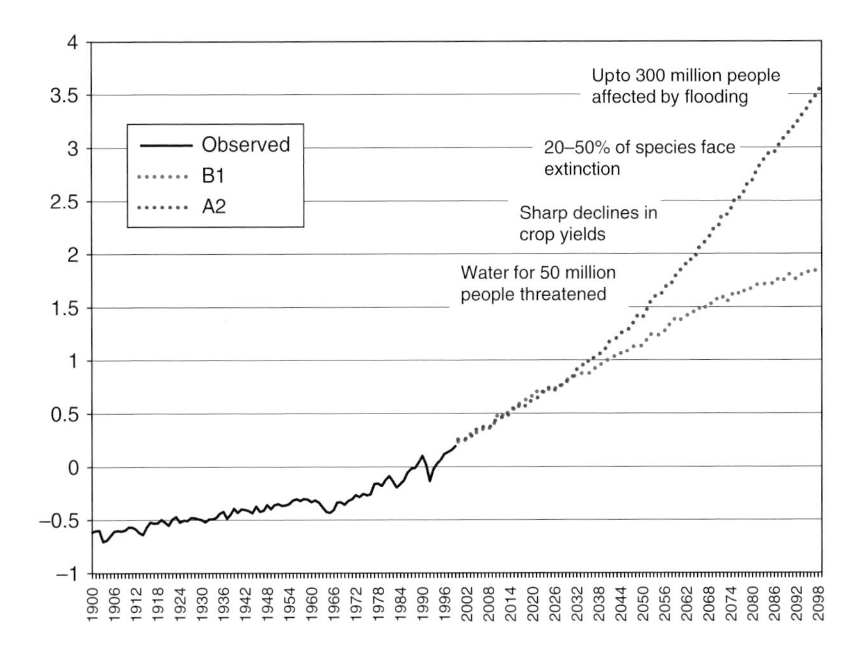

Plate 4. The universe of temperature-rise scenarios, ranging from the most modest to the most dramatic, and a few of the associated climate change impacts.

Source: Intergovernmental Panel on Climate Change, *Impacts, Adaptation and Vulnerability. Contribution of Working Group II to the Fourth Assessment Report of the Intergovernmental Panel on Climate Change*, ed. M.L. Parry, O.F. Canziani, J.P. Palutikof, P.J. van der Linden, and C.E. Hanson (Cambridge: Cambridge University Press, 2007), figure 4.4.

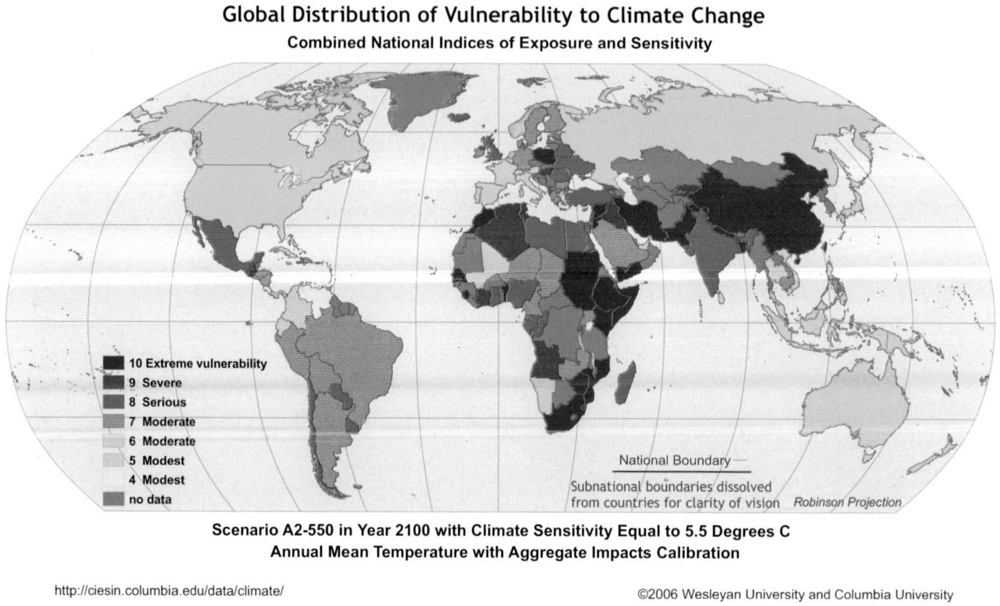

Global Distribution of Vulnerability to Climate Change
Combined National Indices of Exposure and Sensitivity
10 Extreme vulnerability
9 Severe
8 Serious
7 Moderate
6 Moderate
5 Modest
4 Modest
no data
National Boundary
Subnational boundaries dissolved from countries for clarity of vision
Robinson Projection
Scenario A2-550 in Year 2100 with Climate Sensitivity Equal to 5.5 Degrees C
Annual Mean Temperature with Aggregate Impacts Calibration
http://ciesin.columbia.edu/data/climate/
©2006 Wesleyan University and Columbia University

Plate 5. Global vulnerability to climate change, 2100, under a high-emissions scenario.

Source: Columbia University, Center for International Earth Science Information Network (Palisades, NY, 2006).

Plate 6. Kingsway corridor, Burnaby, British Columbia, a street characterized by a sprawling commercial area, few trees, and limited opportunities for transportation other than automobiles.

Source: D. Flanders, in S.R.J. Sheppard, *Visualizing Climate Change: A Guide to Visual Communication of Climate Change and Developing Local Solutions* (Abingdon, UK: Earthscan/Taylor & Francis Group, 2012).

Plate 7. A potential future for the Kingsway corridor, Burnaby, British Columbia, featuring renewable energy, active transportation, local food provision, and public transit.

Source: D. Flanders, in S.R.J. Sheppard, *Visualizing Climate Change: A Guide to Visual Communication of Climate Change and Developing Local Solutions* (Abingdon, UK: Earthscan/Taylor & Francis Group, 2012).

CHAPTER FOUR

Climate Controls: Earth's Reflectivity

MAIN POINTS:

- About 30 percent of incoming solar radiation reflects directly off Earth's surface and off clouds and aerosols in the atmosphere and heads back into space.
- Humans alter Earth's reflectivity through activities such as fossil fuel burning and land-use change.
- In a warming world, feedbacks involving surface albedo (such as ice and vegetation) are primarily amplifying feedbacks. The net result – whether warming or cooling – of feedbacks involving aerosols and cloud albedo is uncertain.
- Mitigation possibilities involving albedo aim to increase the reflectivity of Earth's surface and atmosphere and minimize negative effects.

Have you ever been sunburned under your chin or on the underside of your nostrils? How did that happen? Perhaps you were out in the snow on a sunny day. The Sun's energy reflected off the snow, then back up under your chin and nose, burning your skin just as it would if you stood on your head for long enough and sunburned those parts with energy directly from the Sun. If you shine a flashlight at the ground in the dark, the visible light from the flashlight bounces off the ground and back to your eyes, allowing you to see where you are going. We see the other planets and moons in our solar system because solar energy reflects off their surfaces, then travels across space and through the atmosphere to reach our eyes; they are not hot enough to radiate visible light themselves.

This chapter is about following the photons that reflect via the arrows on the left side of Figure 2.6 (Chapter 5 will follow photons through the right-hand side, which involves the greenhouse effect). Some of the incoming shortwave solar radiation reflects off material in the atmosphere or off Earth's surface, and then heads directly back into space. Pictures of Earth from space record the visible light from the Sun that reflects off Earth's surface. Today, reflection almost immediately turns away about 30 percent of the incoming solar radiation (102 W/m^2 of the 341 W/m^2 that reaches Earth's upper atmosphere).

What does the reflecting? In the atmosphere, clouds and aerosols do the work. At Earth's surface, light-colored ice, snow, and deserts are the most effective, while dark-colored forests and water are less effective. Clean ice reflects away about 80–90 percent of the light hitting it, while ocean water reflects only about 5–10 percent. The reflectivity of a surface is called its *albedo*, defined as the proportion of incident energy that the surface reflects. Thus, the albedo of ice is 0.8–0.9 and the albedo of ocean water is about 0.05–0.1 (Figure 4.1). Earth's overall albedo is about 0.3 – in other words, Earth reflects 30 percent of the incoming solar radiation immediately back into space.

What if Earth's albedo were just a bit higher or lower? How would Earth's overall energy budget change? A change in

Figure 4.1. Approximate Albedo Ranges for Some Major Reflectors in Earth's Climate System

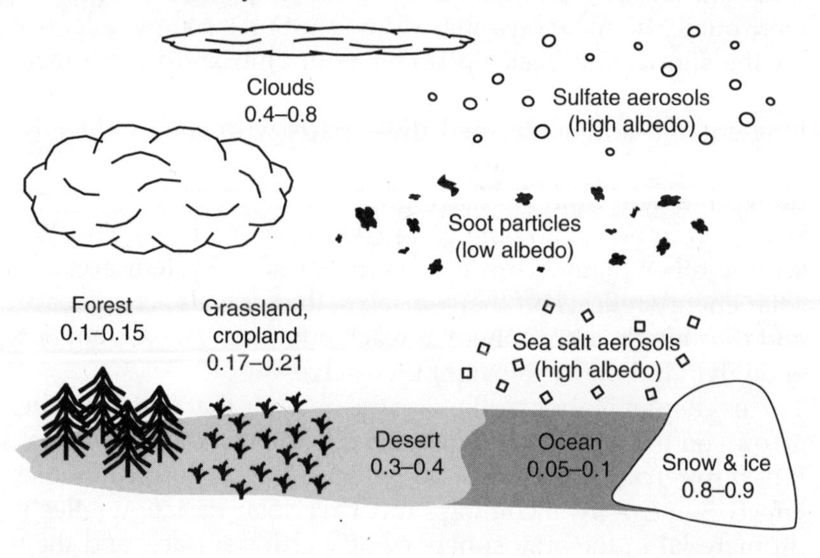

albedo of 0.01 would change the amount of energy absorbed by Earth's system by 341 W/m² × 0.01 = 3.4 W/m². Given a climate sensitivity of 0.75°C per W/m², this translates into about 2.5°C of global average temperature change once the system established equilibrium, if albedo were the only aspect of the climate to change.

Thus, changes in albedo can have substantial consequences for Earth's temperature. In this chapter, we examine the major players in albedo in both the atmosphere and at Earth's surface, discuss both natural and human-induced changes to Earth's albedo, and describe the feedbacks associated with changes in albedo. From a systems perspective, the stocks and flows of reflective materials in Earth's atmosphere and on its surface influence the stock and flow of energy in the Earth climate system. Reflectivity is something humans can and do influence, and therefore is a climate control available to us.

4.1. Natural variability

4.1.1. At Earth's surface: Ice, water, and vegetation

A view of our planet from space, with clouds removed, quickly reveals which parts of Earth's surface have high albedo (because they are light colored) and which have low albedo (because they are dark colored). Look at Plate 1. The oceans, which cover more than 70 percent of the surface, are dark and mostly absorb incoming solar radiation. On land, however, contrasts emerge. Notice the band of dark tropical forest girdling the equator and the lighter-colored bands of desert a bit farther north and south of the equator: the Sahara in northern Africa, the Middle East, dry parts of southern Africa, Australia, the American Southwest, the Gobi Desert; these are dry, light-colored, reflective surfaces. In the northern hemisphere, notice the darker boreal forest sweeping across North America and Eurasia and, at higher latitudes, the lighter tundra, the ice on Greenland, and the sea ice. In the extreme south, notice the very light ice of Antarctica. These alternating bands of higher albedo (deserts, ice) and lower albedo (tropical and temperate forests) occur at different latitudes, a result of the general atmospheric circulation. On average, Earth's surface (the solid or liquid materials at the base of the atmosphere, where we live) reflects about $23\,W/m^2$ of incoming solar radiation. The areas covered by ice, water, and vegetation change naturally over time, so we focus on these three contributors to Earth's surface albedo.

4.1.1.1. Ice

The most intuitive surface reflector is ice. Greenland, Antarctica, mountain glaciers, and sea ice are all good reflectors. The solar energy reflected by these icy surfaces depends on the area covered, the amount of sunlight received (how much energy these surfaces have the opportunity to reflect), and the albedo characteristics of the upper few meters of the ice.

Ice area varies on different time scales and for different types of ice. For its volume, a thin skin of sea ice floating on the ocean can cover a large area – just as a small amount of gold leaf can cover a large dome on an important building. The area covered by this thin sea ice undergoes dramatic seasonal variations, however, as it expands through the winter and melts back in summer. In contrast, the kilometers-thick ice sheets on Greenland and Antarctica expose proportionally less of their volume at the surface, but undergo less dramatic short-term changes in area covered. Changes in ice area are a factor in one of the most important, and most straightforward, climate feedbacks involving surface reflectivity: the ice-albedo feedback (Figure 4.2, panels A and B).

Recall from Chapters 2 and 3 that the ice-albedo feedback is always an amplifying feedback, at all time scales, whether the direction of change is growing ice or melting ice. More ice begets more ice; less ice begets less ice. The initial perturbation that sets the ice-albedo feedback going might be a seasonal change in the distribution of solar energy. For example, summer sunlight in the Arctic initiates sea-ice melt, which is then reinforced by the ice-albedo feedback. Six hundred and fifty million years ago, the ice-albedo feedback helped amplify a cooling trend, plunging Earth into "snowball" conditions in which ice covered most of the planet's surface. During more recent ice ages, subtle changes in Earth's orbit changed the distribution of solar energy and helped initiate this feedback in key locations (see section 3.2.2.4.).

Clearly, ice area matters. It also matters how much sunlight the ice receives, which is largely a function of season and latitude. Since Greenland and Antarctica are located at high latitudes, they receive less incoming solar radiation per square meter than do high-altitude mountain glaciers in the tropics and subtropics, but those mountain glaciers cover a relatively small area. Any ice – sea ice, ice sheets, or mountain glaciers – at latitudes higher than the Arctic and Antarctic Circles, can reflect incoming solar energy only in the sunlit summer. In winter at such latitudes, it is dark, so there is no solar radiation to reflect.

Ice thickness and surface conditions also influence ice albedo. Sunlight penetrates the ice surface and reflects off subsurface

Figure 4.2. Feedback Loops involving Reflectivity at Earth's Surface and in the Atmosphere

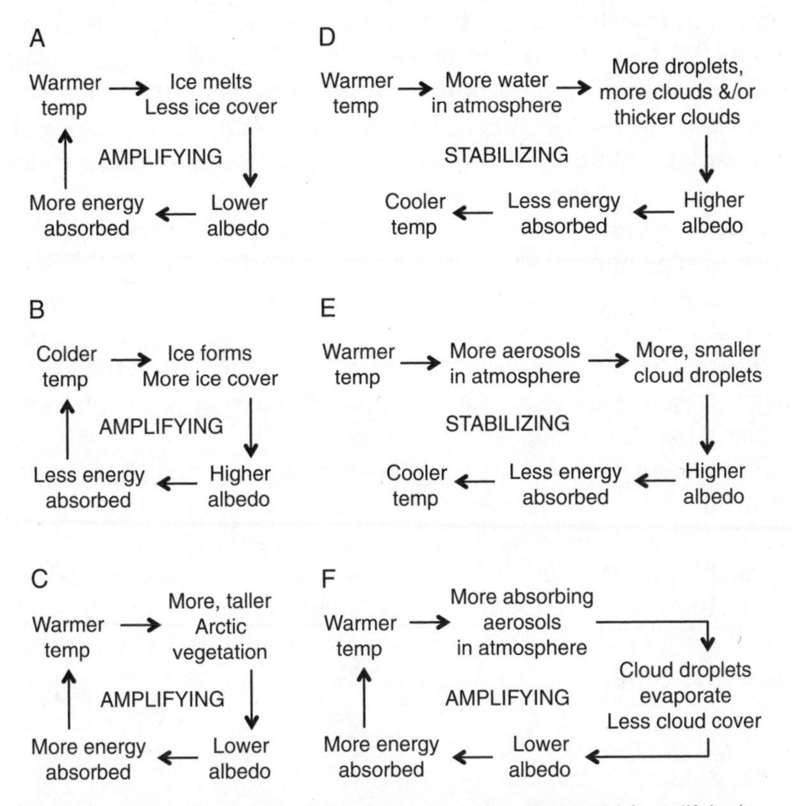

Panel A shows the ice-albedo feedback in a warming world (amplifying). Panel B shows the ice-albedo feedback in a cooling world (amplifying). Panel C shows the Arctic vegetation-albedo feedback in a warming world (amplifying). Panel D shows a possible cloud feedback in a warming world (stabilizing). Panel E shows a possible aerosol-cloud feedback in a warming world (stabilizing). Panel F shows another possible aerosol-cloud feedback in a warming world (amplifying).

ice crystals or, if the ice is thin, gets absorbed by darker surfaces underneath. If you live in a place with snowy winters, you know that the clean snow just after a storm is highly reflective, and the dirty snow that has been around for a while is less reflective. Fresh snow on sea ice, ice sheets, and glaciers brightens those surfaces, while soot, dust, or puddles of melted water on top of the ice darkens them. Dark particles[1] and puddles absorb solar energy and hasten ice melting in the summertime. In the summer of 2012, more than 95 percent of Greenland's surface was under melt conditions simultaneously, decreasing the albedo of a vast area of the ice sheet to about 0.65, such that it absorbed more solar radiation at the time of year when incoming solar radiation was high.[2]

Earth's total albedo due to ice changes with time as the stock of ice changes. As with any system of stocks and flows, the stock of snow and ice over time depends on the balance between inflows (snowfall or sea water freezing) and outflows (melting). Comparison photographs – photos of the same place at two different times – of particular mountain glaciers in the Alps, the Rockies, the Southern Alps of New Zealand, and elsewhere show that the majority have been shrinking in recent decades as outflow (melting) exceeds inflow (snowfall). Greenland and Antarctica are also both losing ice mass.[3] If all of Earth's ice cover melted to expose the land or water underneath, the planet's albedo would decrease, the surface would absorb more solar energy, and, consequently, global temperature would rise.

4.1.1.2. Water and sea level

Conceptually, since the oceans are darker than the continents, rising sea level would cover more of Earth's surface, decreasing Earth's total albedo. In turn, Earth would absorb more radiation, global temperature would go up, and sea level would rise further – another amplifying feedback. Thus, global albedo can vary as the boundaries between continents and oceans change. As discussed in Chapter 3, during the most recent ice age, sea level was about 120 meters lower than it is today, because much more of the water on Earth was locked up in giant ice sheets on land. Where New York City is located would have been more than 150 kilometers from the coast, and the extra land exposed at that time would have been more reflective than the dark water covering that area today. If the ice on Greenland and Antarctica were to melt away completely, returning the water to the ocean, sea level would rise about another 70 meters, completely submerging buildings in New York lower than about eighteen stories. The magnitude of an albedo change arising from water taking over land is fairly small on human time scales, simply because it takes a lot of sea-level rise to cover large areas of continent. If sea level were to rise 10 m (a lot for human time scales), less than 1 percent of Earth's surface area would switch from continent to ocean.[4]

The primary contributors to sea-level rise today are thermal expansion and ice melting from glaciers and ice sheets. Water expands when it warms, meaning that the same number of water molecules takes up more volume. Even without adding any water to the oceans, sea level rises if the ocean warms, even in situations (in Earth's past) when there were no glaciers and ice sheets to melt and add more water to the ocean basins. Globally, sea level has been rising about 3.2 ± 0.4 mm/year in the past two decades (this notation means that the likely "true" rate is between 2.8 and 3.6 mm/year, with a best estimate of 3.2 mm/year), and has averaged about 1.7 ± 0.2 mm/year since 1900,[5] due to both the thermal expansion of ocean water and the contribution of melting ice.

On longer, tectonic time scales, sea level can change Earth's surface albedo if rates of sea-floor volcanism change. When rates of volcanism are high, underwater volcanoes expand in volume, taking up more space on the bottom of the ocean basins and pushing sea level higher. When rates of volcanism are lower, the volcanoes are cooler and smaller in volume, taking up less space. For time scales relevant to human mitigation and adaptation, however, changes in rates of sea-floor volcanism do not matter much.

Although its albedo effect might be small on time scales of most interest to us, sea-level rise does produce darker surfaces, and thus promotes warming. From a human perspective, however, sea-level rise is a major challenge for reasons other than albedo. It has serious potential to destabilize human societies, as coastal dwellers move inland in advance of rising seas, as some have already been forced to do. It also causes costly damage to infrastructure, loss of livelihoods, and public health problems as sanitation systems are overwhelmed during floods.

4.1.1.3. Vegetation

What would happen to Earth's albedo if forests expanded across all the deserts of the world? What about the reverse? Deserts, regions in the rain shadow of mountain ranges, and sparsely vegetated areas above the timberline and poleward of the treeline typically have high albedo. As Figure 4.1 shows, dark forests have low albedo; grasslands have intermediate albedo. Ignoring any feedbacks involving the hydrological cycle or other consequences, if forests took over deserts, albedo would decrease and Earth's surface would absorb more incoming solar radiation, warming the planet. If deserts took over all the forests, Earth's surface would get brighter, reflect more, and cool.[6]

Over time, the natural ground cover does change. Long-term changes – on time scales of tens of millions of years – happen because continents move as a result of plate tectonics. If, for example, most of the continental area were concentrated in the tropics, much of the land surface would be dark tropical forest. If the continents were concentrated in the dry subtropics, or at the poles, they likely would be more reflective. The locations of large mountain ranges, driven by slow, tectonic processes, also influence where rain falls and vegetation grows.

Desert areas also grow and shrink. During the most recent ice age, the continents were generally drier than they are today; deserts such as the Gobi and the Sahara were larger and tropical forests were smaller, helping the glacial world stay cool. Within the relatively warm past 10,000 years, precipitation changes have helped the boundary between the dry, light-colored Sahara desert and the greener, darker Sahel to its south fluctuate north and south, changing the albedo of that part of the world over time.

Forests play multiple roles in the climate system, but in terms of their albedo, they are poor reflectors – darker than adjacent grasslands, bare rock, or ice. They can block sunlight from reflective surfaces underneath, as when the branches of tall trees intercept solar radiation before it can reflect off white snow on the ground in winter. But indirectly, forests can also increase a region's albedo, particularly in the tropics, through evapotranspiration. Evapotranspiration can promote the formation of clouds, one of the most important atmospheric reflectors (see section 4.1.2.2.), and cool the surface through latent heat transfer, unrelated to albedo. All things considered, tropical forests appear to promote a net cooling, while the net effects of mid-latitude forests on temperature is uncertain.[7]

Warming temperatures can initiate important feedbacks involving vegetation and albedo. In response to warming temperatures in the high Arctic in the past few decades, vegetation cover has increased as taller shrubs encroach northward into territory previously dominated by shorter tundra vegetation. The taller shrubs stick above the snow and decrease albedo. The Arctic growing season starts earlier and ends later than it used to, providing more days per year for vegetation growth. All these changes, initiated by warming, add up to greener, darker-colored tundra, with lower albedo, which enhances warming – another amplifying feedback (Figure 4.2, panel C).

In summary, the reflectivity of Earth's surface responds to, and influences, global temperatures. If we look back in Earth's history, changes in surface albedo due to feedbacks involving ice area, sea level, and vegetation have played important roles, mostly amplifying temperature changes. They were likely responsible for about half the global temperature change during transitions between the naturally occurring cold glacial and warm interglacial periods of the past million years.[8] Surface albedo will be an important contributor to future temperature changes as well.

4.1.2. In the atmosphere: Aerosols and clouds

Throw a handful of glitter in the air, and you will see light reflecting off hundreds of little mirror-like particles. Spray paint a wall, and you will add tiny droplets of paint to the atmosphere in addition to coloring the wall. Boil water, and you will see the resulting water vapor condense into droplets as it cools in the air above the pot. Your glitter, spray paint, and steam droplets are analogous to the reflective materials suspended in Earth's atmosphere. Aerosols and clouds in the atmosphere scatter and reflect incoming solar radiation, sending about 79 W/m^2 of the solar energy back into space – more energy than reflects off Earth's surface. Like reflective materials at Earth's surface, atmospheric reflectors decrease the amount of energy Earth absorbs.

Glitter particles might seem small in size, but compared with the tiny particles and droplets in the atmosphere that do the job of reflecting incoming solar radiation, glitter is huge. If you stopped throwing glitter, you could watch your atmosphere become glitter free as the glitter settled out. Tiny aerosols – the "large" end of the size range for aerosols is about 100 microns, or 1/10th of a millimeter – are constantly added and removed from the atmosphere, with the aerosol stock dependent on the balance of inflows and outflows over time. Some aerosols cycle through the atmosphere quickly and others more slowly, but in general aerosol lifetimes are years or less. If a source (inflow) of aerosols were cut off or decreased, the stock would decrease fairly quickly.

The natural atmospheric glitter is mostly the water droplets in clouds, sea salt, dust, sulfur compounds, volcanic particles, soot, and other organic particles. Of these, clouds are responsible for the largest portion of energy reflection although they interact in important ways with aerosols. Aerosols can change the reflective properties of clouds, while clouds can influence the concentration of aerosols.

4.1.2.1. Aerosols

What are the natural glitter throwers? The primary culprits are volcanoes, wind, ocean waves, forest fires, and organisms.[9] Volcanoes intermittently spew ash and gas, wind picks up dust from deserts and dry soil, wind-driven ocean waves spray sea salt into the atmosphere, forest fires produce smoke and ash, and living organisms release pollen and sulfur products.

Large, powerfully explosive volcanic eruptions can inject gases and particulates all the way into the stratosphere, higher than about 10 km above Earth's surface. Volcanic ash is the most visually spectacular part of an explosive eruption, but it is composed of fairly large particles, and so settles out of the atmosphere quickly. Most of the ash lands relatively close to the volcano, though fine particles from large eruptions can be carried downwind thousands of kilometers. Some fine particles that make it into the stratosphere can persist for a few months and a trip or two around the globe. A cloud of ash, in contrast, might last a few weeks, possibly disrupting air travel, but it has a relatively brief influence on Earth's reflectivity.

Gases emitted by volcanoes, particularly sulfur dioxide (SO_2), are another story. SO_2 gas undergoes fast chemical reactions that form sulfuric acid (H_2SO_4). The sulfuric acid, plus water, condenses into tiny droplets, and these aerosols are responsible for the bulk of the reflection of incoming solar radiation due to volcanic activity. High up in the stratosphere, these droplets can persist for several years. A large eruption can have serious consequences for Earth's climate in the years following. The "year without a summer," 1816, that followed the massive eruption of Mount Tambora in Indonesia in 1815 resulted from volcanic aerosols reflecting so much incoming solar radiation that crops failed, causing famine all around the globe.

Thus, large eruptions can cause a year or more of global cooling by changing the albedo of Earth's atmosphere. In the recent past, large, well-observed eruptions have provided natural laboratories to estimate the effects of radiative forcing on global temperatures. But is there one large eruption per century? three? Along with variability in solar output (Chapter 3), the background rate of volcanic activity is one of the most important natural contributors to global climate change, and must be taken into account when attempting to partition changes in global temperature between natural and human causes. Volcanoes also provide the inspiration for some ideas about responding to the effects of greenhouse warming (see section 4.3.).

Other aerosol sources are much more consistent and continual than volcanoes. Wind picks up particles of dust from deserts and carries them, sometimes thousands of kilometers, through the atmosphere. Dust from the Sahara Desert regularly reaches the Caribbean; you can find spectacular satellite images of dust storms on the Internet. Dust both reflects and absorbs radiation, thus its net effect – cooling or heating – depends on conditions such as the composition of the dust, its altitude, whether it is above or below clouds, and the character of the surface underneath. As well, wind-driven waves breaking anywhere on the ocean's surface throw tiny water droplets and salt into the atmosphere that act as excellent little reflectors. These are short lived in the atmosphere, but are produced at a higher rate than any of the other natural aerosols, and are constantly replenished.

The last natural aerosol sources we will discuss here come from biological materials. Fires, in any vegetated environment (forests, grasslands, brush, tundra) release combustion products into the atmosphere, which are then transported by the wind. Find a satellite image of forest fires and you will see plumes of smoke stretching downwind. Soot from fires is a lousy reflector, and absorbs radiation better than it reflects. But other biological sources help increase albedo – for example, marine phytoplankton produce and emit natural dimethyl sulfide into the atmosphere, where it oxidizes to sulfate and contributes to the natural, and reflective, sulfate aerosols found there.

Most aerosols ultimately get removed from the atmosphere gravitationally, with help from various chemical and physical processes that promote coagulation, or incorporation into larger entities, such as raindrops. Aerosols between 1 and 100 microns in size, such as volcanic ash and dust, tend to fall out of the atmosphere within hours or days. Particles transported by the wind close to Earth's surface can run into an obstacle (a building, a tree, an ice sheet) and stick to it, and thus are removed from the atmosphere. Particles between 0.1 and 1 micron tend to get removed with rain or snow, as aerosols provide nuclei around which water droplets or ice crystals form. They then precipitate to Earth's surface by tagging along in the hydrologic cycle. Even if they are not the nucleus of a water droplet, aerosols can get scavenged by falling precipitation, which is why the air seems "cleaner" after a rainstorm. Very tiny aerosols can stay in the atmosphere for months to years. Eventually, they coagulate with one another into larger aerosols, or undergo chemical reactions, and are also removed.

4.1.2.2. Clouds

On a foggy day, it can be hard to see across the street, because most of the visible light bouncing around does not have a clear path from your neighbor's house to your eyes, but instead it encounters water droplets that reflect and scatter it. Clouds – made of water droplets – are great reflectors of incoming shortwave solar radiation. Even though the stock of water in clouds is tiny compared with reservoirs of water in, say, the oceans, clouds are the largest contributor to Earth's overall albedo, accounting for about half the energy reflected.

Condensation, evaporation, and precipitation are the key processes for flows of water in and out of clouds. Water flows into clouds when atmospheric water vapor condenses into droplets or ice crystals. Outflows occur when those droplets or ice crystals either precipitate to Earth's surface or evaporate again within the atmosphere, as when clouds "burn off." In general, water cycles in and out of the atmosphere quickly: the average time spent by a water molecule in the atmosphere is about ten days – even less for water in a cloud.

The amount of solar radiation clouds reflect depends on the area clouds cover, their thickness, the sizes of water droplets or ice crystals they contain, and the concentrations of those water droplets or ice crystals. Any unedited satellite image of Earth will show lots of area with some degree of cloud cover. You can imagine that a planet entirely covered by clouds would allow very little solar radiation to make it through to its surface. Thick clouds provide more opportunities than thin clouds for solar radiation to be reflected and, in general, thicker, more reflective clouds form at lower and middle altitudes. Thin, wispy clouds, often in the upper atmosphere, are poorer reflectors. Small droplets and more droplets per unit volume increase the brightness of clouds, making them more reflective. Thus, more area covered, thicker clouds, smaller droplets, and higher droplet concentrations all increase cloud albedo. Variations in these parameters change the overall amount of incoming solar radiation clouds reflect.

Feedbacks involving cloud cover and albedo are among the least well understood aspects of Earth's climate system. Theoretically, if warming temperatures result in greater cloud cover, and those clouds reflect more incoming solar radiation, then cloud reflectivity is a stabilizing feedback in the climate system, counteracting the warming (Figure 4.2, panel D). But if warming temperatures result in less cloud cover, cloud reflectivity is part of an amplifying feedback. And clouds at different latitudes might respond differently to a warming world. In the Arctic, as sea ice melts (decreasing the albedo due to ice cover), more water is exposed to the atmosphere, promoting evaporation, promoting cloud formation, and increasing albedo again, but now the reflection is off clouds rather than off sea ice. The albedo effect of forming those new Arctic clouds is a stabilizing feedback. At the edges of the subtropics, in contrast, warmer conditions might evaporate more clouds, decreasing cloud cover. The challenge is to figure out the magnitude and direction of changes in cloud albedo under the various conditions in different locations, to estimate net global effects.

Beyond the area covered by clouds, determining cloud feedbacks is complicated by details such as whether the clouds that form are thick or thin, or have small or large droplets. It turns out that aerosols, our other main contributor to atmospheric albedo, are intimately involved with changes to the reflective properties of clouds.[10] If aerosols and clouds were independent of one another and did not interact, perhaps the climate effects of these two key atmospheric constituents would be a bit more straightforward. But they do interact, giving rise to some important feedbacks in the climate system involving albedo.

Aerosols provide tiny surfaces for water vapor to condense onto in the atmosphere, helping cloud droplets form. If lots of aerosols are present, the available water will condense into many small droplets as opposed to fewer large droplets. For a given amount of liquid water in a fixed volume of cloud, the cloud will have more total droplets if those that form are small, and fewer total droplets if those that form are large. Thus, the presence of aerosols mediates both the droplet size and droplet concentration – the number of droplets per volume – in clouds.[11] Both smaller droplets and more droplets per cubic centimeter increase cloud brightness; thus, more aerosols can increase cloud albedo (see Figure 4.2, panel E).

Subsequent effects follow from smaller droplet sizes. Smaller droplets take longer to form raindrops or ice crystals that will precipitate gravitationally to Earth's surface. This delay can help increase the water content in clouds (as condensation continues), or make clouds last longer, or increase cloud thickness. Thicker clouds containing more water in small droplets are more reflective, and if a cloud sticks around longer, it has more time to reflect incoming solar radiation. These additional effects of aerosols also increase cloud albedo and enhance cooling.[12]

Not all effects of aerosols on clouds increase cloud albedo. Aerosols do not only scatter and reflect incoming solar radiation; they can also absorb it, particularly dark aerosols like soot. By absorbing solar radiation, the now-warmer aerosols help heat the surrounding air, which energizes the water molecules. More water molecules then evaporate from the liquid phase into the vapor phase,[13] a process particularly effective if the initial cloud droplets are small. This warming of clouds by aerosol absorption of energy reduces cloud cover (the clouds burn off), reduces cloud albedo, and helps more solar radiation reach Earth's surface (Figure 4.2, panel F).

Feedbacks involving clouds in a warming world are complicated. The magnitudes of the different effects are difficult to measure and disentangle, and both amplifying and stabilizing feedbacks are operating. It appears likely, however, that, in a warming world, the net cloud-related feedbacks are overall amplifying.[14] In a warmer world, there is more water in the atmosphere than in a colder world. That water might be present as vapor (a greenhouse gas that absorbs longwave radiation), rather than as reflective liquid droplets or solid ice crystals in clouds. And dry regions with clear skies might expand in a warming world, decreasing global albedo and increasing absorption of solar radiation at Earth's surface.[15] Like reflective volcanic aerosols, features of clouds and cloud albedo have inspired various geoengineering ideas to promote cloud cover or encourage smaller cloud droplets to counteract greenhouse warming (see section 4.3.).

4.2. Anthropogenic variability

4.2.1. Land-use changes

Humans have been altering the albedo of Earth's surface for a long time. We clear forested land for agriculture, mining, or urban and suburban development. As urban populations grow, we convert agricultural land to buildings, roads, and other human infrastructure. We set fires to alter vegetation cover; our domesticated animals change the vegetation on grazed landscapes, sometimes promoting desertification. We also abandon agricultural lands, on which forests naturally regrow, and we deliberately plant trees in areas that were not recently naturally forested. These are all examples of more direct ways in which we alter surface albedo. Consider, however, that many of the "natural" surface albedo changes in the past century or so are also responses to warming temperatures attributed to human activities.

Many of the large-scale surface albedo changes due to human activities involve vegetation, particularly forest cover. Evidence of our clearing forested land for agriculture extends back at least eight thousand years, based on changes in the makeup of pollen preserved in lake sediments and on data showing increased rates of erosion. Large areas of East Asia, Europe, and North America have been transformed from forest to agricultural uses since that time. Recently, tropical forests have undergone the fastest rates of conversion from forest to crops and pasture. Since cropland and pasture are lighter in color than forest, the expansion of agriculture into previously forested lands has increased Earth's surface albedo. Degradation of natural grasslands through grazing also increases surface albedo if vegetation cover decreases and desert-like conditions encroach on former grasslands.

We also sometimes increase vegetation cover, both inadvertently and deliberately. For example, large-scale reforestation occurred in naturally forested eastern North America during the twentieth century after farmland was abandoned in favor of concentrating agricultural efforts in the prairies of central North America. The regrowth of trees decreased surface albedo again. In that case, forests regenerated on their own after disturbance, but sometimes we also deliberately plant trees in places that are not naturally forested, which also decreases surface albedo.

Why would we deliberately darken Earth's surface by planting trees, if today's problem is one of warming? As you might have guessed, albedo is not the full story. Trees and vegetation play other roles in the climate system, and the combination of all effects is what matters, not just surface albedo. As we mentioned in section 4.1.1.3., evapotranspiration by plants cools the surface, and the resulting atmospheric water vapor can generate reflective clouds in forested areas. Vegetation removes carbon from the atmosphere through photosynthesis and stores it in plant biomass, decreasing the greenhouse effect. Tropical deforestation, even though it increases surface albedo, might result in net warming if a loss of cooling from plant evapotranspiration exceeds the increased cooling gained from surface albedo.

In addition, human populations are increasingly urban, and land occupied by urban landscapes continues to expand. The change in albedo due to urbanization depends on what is built and what is replaced. Planting trees for shade and aesthetics in urban areas surrounded by dry landscapes can decrease the albedo of the altered area. Conversely, replacing forested land with buildings typically raises albedo. Conversion of agricultural land to urban or suburban development typically lowers albedo.

The net effect of human land-use change on Earth's surface albedo depends on the area converted to a new type of surface and on the difference between the albedo of the new and former surface. Since the 1970s, satellite data have been the best source of information about surface albedo changes, as satellites can generate repeated images of nearly the whole of Earth's surface. Prior to the satellite era, records of land-use change rely on historical land surveys, and before land surveys, on reconstructions from the geologic record. For the industrial era, most studies conclude that human land-use changes that have altered surface albedo have changed the energy Earth absorbs by about $-0.15\,W/m^2$, resulting in a net cooling.[16]

4.2.2. Anthropogenic aerosols

Anthropogenic aerosols, primarily from burning fossil fuels and biomass, probably first gained widespread attention because of their effects on air quality and human and ecosystem health, rather than for their role in global climate. The products of coal burning combined with fog in London in the 1950s produced such poor visibility at times that someone had to walk in front of the city buses to guide the driver. Air quality was so poor for a week in December 1952 that thousands more people died and became ill than usual, prompting government regulation to control emissions. In the United States, decades of research into air pollution, spurred by concerns about human health and acid rain, led to the 1970 Clean Air Act and subsequent updates. Government regulations resulted in decreased emissions of sulfate and particulate matter from coal-burning power plants, higher standards for vehicle emissions, and other controls on anthropogenic aerosols. From an air-quality perspective, reducing anthropogenic aerosols has clear health benefits.

From an albedo perspective, some anthropogenic emissions produce highly reflective aerosols, increasing global albedo and promoting cooling. Of reflective anthropogenic aerosols, sulfates are most prominent. Our main sulfate emissions are from burning coal, particularly high-sulfur coal. Just as with natural sulfates released from volcanoes or marine phytoplankton, sulfates from coal form tiny droplets of sulfuric acid in the atmosphere and reflect incoming solar radiation. Coal burning for human energy consumption has increased in recent decades,[17] but due to clean air legislation in some countries, global inflows of coal-generated aerosols to the atmosphere have decreased.[18] Sulfate aerosols today are regionally concentrated in areas with high coal use and few regulations on sulfate emissions. Coal is the largest stock of conventional fossil fuel, and if we burn it without scrubbing out the SO_2 at the source, we can expect continued sulfate emissions.

These anthropogenic sulfates are one edge of a double-edged sword. Burning fossil fuels produces both reflective sulfates and CO_2, the most critical greenhouse gas today. The CO_2 causes heating, and the sulfates partially offset the heating. Although technology to capture SO_2 is used at many coal-fired power plants, large-scale CO_2 capture remains rare. Thus, in our current mode of operation, we preferentially remove a cooling agent and simultaneously emit a warming agent. Even if we did not invest in SO_2-removing technology, sulfate aerosols have a much shorter lifetime in the atmosphere than does CO_2, so if we stopped burning coal today, the reflective aerosols would quickly be gone, while the long-lived greenhouse gases would remain.

Other anthropogenic aerosols, such as soot, both absorb radiation while in the atmosphere and darken the surfaces on which they land, promoting warming. Burning biomass and fossil fuels (particularly coal and diesel fuel) produces dark soot particles. While in the atmosphere, these particles absorb incoming solar radiation, warming the atmosphere and interacting with clouds (section 4.1.2.2.). But perhaps more important, the soot lands somewhere downwind, sometimes far from the source, altering the albedo of the surface where it lands. Anthropogenic soot landing on white, highly reflective ice can darken the surface enough to have a globally significant warming effect.[19] Additional subsequent effects are increased melting of the now-darker ice, increased transfer of water to the oceans, and sea-level rise.

On a basic level, human activities simply add additional particles into the atmosphere. These have their own reflective properties, but also influence the albedo characteristics of clouds. Under natural, fairly low aerosol concentrations, clouds have fairly large droplets and low droplet concentrations. Under high aerosol conditions (typically due to human activities), aerosol concentrations can triple, which provides more potential cloud condensation nuclei, decreases droplet sizes, and increases droplet concentrations, but can also promote evaporation.[20] The question then becomes whether the net anthropogenic aerosol effect on clouds is one of cooling or warming. In the 2007 IPCC Fourth Assessment Report,[21] the overall effect was estimated to be cooling, but with large uncertainty. The aerosol-cloud-albedo story might be different by the time you read this.

Anthropogenic aerosols are complicated. Their overall effects – the sum of direct reflection of incoming solar radiation, absorption of solar radiation, subsequent effects on cloud albedo and land surface albedo – are among the primary areas of uncertainty in our current understanding of global climate. Nonetheless, deliberate emissions of reflective aerosols are considered one of our mitigation options.

4.3. Response strategies

In Chapter 3, we briefly discussed the limited (and expensive) options for altering the amount of solar energy arriving at Earth's upper atmosphere. That, plus climate change response options involving reflectivity, are often grouped as "solar radiation management" strategies. Such strategies using reflectivity (after the solar radiation has arrived) involve efforts to alter the overall reflective properties of the atmosphere or Earth's surface. In our current situation, this means trying to make the atmosphere and Earth's surface more reflective, to counter warming, which is primarily occurring because of greenhouse gases. All of the strategies discussed in this section involve trade-offs among cost (both initial investment and maintenance), effectiveness (how much they would actually alter the W/m^2 reflected), and safety (the reversibility of the approach, and its potential negative effect on biological communities, atmospheric circulation, the hydrologic cycle, or other aspects of the Earth system).[22]

What can we do to increase the albedo of Earth's surface? One simple approach is to paint roofs and roads white. Although this would increase albedo in areas with human infrastructure and would have few negative effects on ecosystems, this approach would be fairly expensive, and the area of Earth's surface covered with buildings and roads simply is not large enough to make much difference to global albedo. Other types of land surfaces encompass larger proportions of the total, and thus have better potential. For example, since humans control vegetation cover across large swaths of agricultural land, if we planted more reflective crops, we could increase Earth's surface albedo. But would more reflective crops provide the same agricultural yields, and how might particular large-scale crop changes alter the hydrologic cycle? A third way to change surface albedo would be to cover sunny desert regions with a synthetic reflective cover. Although this is attractive, since we do not rely much on deserts for our sustenance, the costs of maintaining (or removing) such a cover and the unknown consequences for biological ecosystems both in the deserts and downwind make this a less desirable option.[23] In addition to lightening surfaces deliberately, we could decrease the rate at which we darken them. Decreasing soot emissions, over time, would lighten the natural surfaces of ice sheets as new snow covers previous soot deposits.

To increase the reflectivity of the atmosphere, we could add reflective particles or increase the reflectivity of clouds. We could imitate volcanoes by injecting sulfates into the stratosphere, where they would last a few years, reflecting incoming sunlight, and cooling the planet. This approach could be highly effective, but it likely would have undesirable effects on stratospheric chemistry, atmospheric circulation, and biological productivity. We could also increase sulfate emissions to the troposphere, by burning coal without scrubbing out the sulfates, but at the cost of human health problems, acid rain, and potential unknown changes in regional climate. This technique would be quickly reversible (just stop adding sulfates), but if we came to rely on it as a primary climate change response strategy to counter the warming effect of greenhouse gases, we would be committed long term. Deliberately adding reflective sulfates to the atmosphere might be both effective and cheap, but with low margins of safety.

Various techniques have been proposed to enhance the availability of cloud condensation nuclei, and thus increase cloud cover, decrease droplet size, or increase droplet concentration. Wind-powered autonomous machines could increase sea spray in remote parts of the oceans. Stimulation of marine biological productivity could increase release of dimethyl sulfide into the atmosphere. These possibilities have poorly constrained consequences for marine ecosystems, but would be fairly quickly reversible.

Most albedo-based responses to climate change aim to offset the warming induced by greenhouse gases.[24] They are analogous, however, to treating symptoms – like taking a second drug to counter the negative effects of the first, or lowering a fever without curing the disease. The primary culprit in recent global temperature rise is an increase in the greenhouse effect – that is, we have a greenhouse gas problem. Strategies for mitigation that involve reducing atmospheric greenhouse gas concentrations are crucial if we want to minimize future global temperature rise. Albedo-based strategies could help, but the longer we put off addressing the root problem, the longer we might be committed to albedo treatments, many with poorly known side effects.

CHAPTER FIVE

Climate Controls: The Greenhouse Effect

MAIN POINTS:

- Greenhouse gases in the atmosphere absorb particular wavelengths of radiation emitted by Earth's surface, then re-emit the radiation in all directions, slowing the energy's passage back into outer space and warming Earth.
- Long-lived CO_2 is the primary gas that drives changes in the greenhouse effect.
- The stock of CO_2 in the atmosphere depends on historical relative magnitudes of inflow and outflow to and from the atmosphere as the atmosphere exchanges CO_2 with other carbon reservoirs.
- Key stabilizing feedbacks related to the greenhouse effect are the longwave radiation feedback and the lapse rate feedback. Key amplifying feedbacks are the water vapor feedback and longwave cloud feedback.
- Human actions perturb the carbon cycle, particularly by increasing the inflow of CO_2 to the atmosphere. As a result, atmospheric CO_2 is rising extraordinarily quickly compared with records from the geological past.

Venus's greenhouse effect makes that planet insufferably hot, while Mars's greenhouse effect is so small that the planet's average surface temperature is similar to that in the middle of Antarctica. Lucky Earth has a greenhouse effect that warms the planet to a hospitable range of temperatures. Without the greenhouse effect, but with our present albedo, Earth would have an average temperature similar to Winnipeg in winter (–18°C). It would be colder than that, actually, because that cooler Earth would have more ice cover and thus higher albedo than Earth has today. Instead of that icy world, we have a comfortable planet, suitable for a huge range of life forms.

Of the three basic factors controlling energy flows in the climate system, the most important today is the greenhouse effect. Its magnitude is changing rapidly, the consequences are long lived and the agent of change is us. We humans have our hands gripped firmly on the spigot that controls the inflow of greenhouse gases to the atmosphere. Over our industrial history, we have freely spun that valve open ever wider (with a few deceleration episodes). We now primarily control atmospheric greenhouse gas concentrations through our manipulation of inflows. So far, for outflows, we largely rely on plants, soil, and the oceans. Our focus on inflows is not deliberate planning from a systems-thinking perspective, but, rather, reflects our interest in accessible energy sources and our insatiable demand for goods and services. Today, burning fossil fuels, which adds carbon dioxide (CO_2) to the atmosphere, accounts for more than 80 percent of our energy supply.[1] Over the industrial era, our use of fossil carbon for energy has grown from virtually zero to about 9 GtonC/year in 2011.[2] In total during that time, we have eaten up about half of our total carbon "budget" – the amount of carbon emissions we must not exceed if we are to have a reasonable chance of keeping warming to less than 2°C above pre-industrial temperatures.[3] Today, if you travel by motorized vehicle, use your electrically powered gadgets, or turn on the air conditioning, chances are good that the energy you use releases CO_2 to the atmosphere – you even make a tiny contribution to Earth's greenhouse effect while you type.

In this chapter, we follow the photons through the energy flow pathways on the right-hand side of Figure 2.6. What happens to the energy that radiates from Earth's surface? What physical processes act to keep energy in Earth's system for longer than it would stick around on a planet without a greenhouse effect? What are some of the important feedbacks, both amplifying and stabilizing, involving the greenhouse effect? How do our actions influence the greenhouse effect, and how do we know? Because our manipulation of carbon flows is among the most crucial topics for mitigation, we devote two chapters to the issue: Chapter 5 deals primarily with the science, and Chapter 6 delves into potential solutions.

5.1. How does the greenhouse effect work?

5.1.1. Characteristics of a good greenhouse gas

Fundamentally, a greenhouse gas is one that can absorb and re-emit energy within the range of wavelengths of radiation given off by Earth's surface. Recall from section 3.1.1. that, because of its average temperature of about 288 K, Earth emits longwave radiation, in the infrared part of the electromagnetic spectrum.[4] Greenhouse gases are those that have the right chemical structure to interact with this infrared radiation on its way back to space. Gases that do not interact with radiation in this wavelength range are not greenhouse gases. Beyond this one crucial criterion, the effectiveness of any particular gas to help increase Earth's temperature depends on (1) the particulars of the wavelengths it can absorb and re-emit; (2) its concentration in the atmosphere; and (3) its lifetime in the atmosphere.

Greenhouse gases are picky. CO_2, for example, is an excellent absorber (and re-emitter) of 15-micron wavelength radiation. Methane (CH_4) and nitrous oxide (N_2O) both like wavelengths near 7.5 microns. Water (H_2O) has the broadest range of capabilities, and interacts with many different wavelengths of radiation, across the infrared spectrum (Figure 5.1). The gases with

Figure 5.1. Absorption and Re-emission of Radiation by Greenhouse Gases

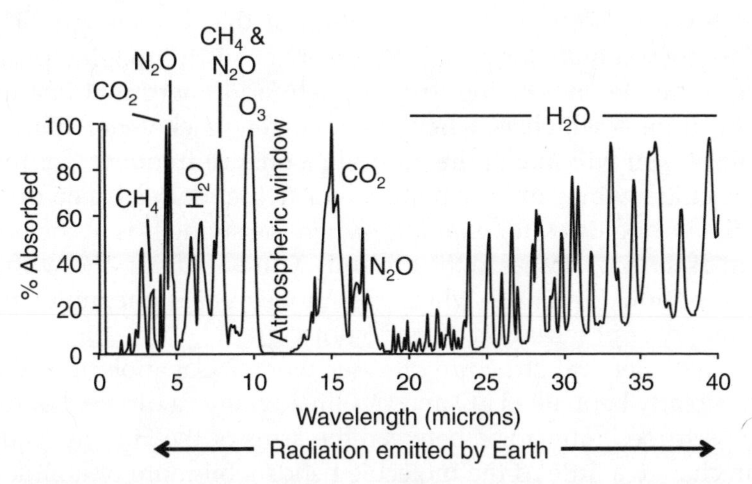

Greenhouse gases absorb and re-emit radiation at particular wavelengths. All the absorption bands to the right of about 19 microns are due to water vapor. The higher the "percent absorbed" value, the better the gas is at absorbing that wavelength. The "atmospheric window" is the wavelength range between about 10 and 12 microns. Radiation at these wavelengths passes through the atmosphere without interacting with greenhouse gases.

preferences for wavelengths close to the peak of Earth's radiation spectrum (Figure 3.3) have the most opportunity to absorb and re-emit infrared radiation. For example, CO_2 and ozone (O_3) are both well situated near Earth's emission peak. Gases with absorption preferences farther towards the edges of Earth's radiation range do not have as many absorption opportunities.

Why do gases have such particular preferences? Gas molecules are not rigid; they can bend back and forth or stretch side to side, producing vibrations. Each type of gas vibrates at preferred frequencies, based on its number of atoms, bond strengths, and geometry. Some of these vibrations produce asymmetrical changes in the arrangement of positive and negative charges in the molecule, creating a dipole. When a greenhouse gas molecule encounters infrared radiation with frequency characteristics that match the frequency at which the molecule naturally vibrates and produces a dipole, the molecule will absorb and emit radiation at that wavelength. The most common analogy for such resonance is to imagine pushing someone on a swing. If you push (add energy) at *just* the right time – which is when the swinger reaches her highest point – you will match the natural resonance frequency of that particular swing, and the person will swing higher. If you push at the wrong time (for example, when the swinger is at the bottom of the arc, closest to the ground), you will damp the swinger's motion. Similarly, when greenhouse gases encounter the wrong type of radiation, they do not vibrate.

What chemical structure of gases works? Gas molecules that are already bent, like H_2O and O_3, already have a dipole that can be perturbed. The angle between the arms of the H_2O molecule can change a little as the molecule bends a bit more or a bit less (Figure 5.2). Or, the distance between the central oxygen atom and the hydrogen atoms out on the water molecule's limbs can increase or decrease as the molecule stretches a bit. A molecule does not have to be bent permanently to be a greenhouse gas: CO_2 has a linear, symmetrical structure, with the carbon atom in the middle of the two oxygen atoms. But because CO_2 has three atoms, it can bend a little, forming a slight V, and it can stretch asymmetrically such that one of the oxygen atoms gets a little

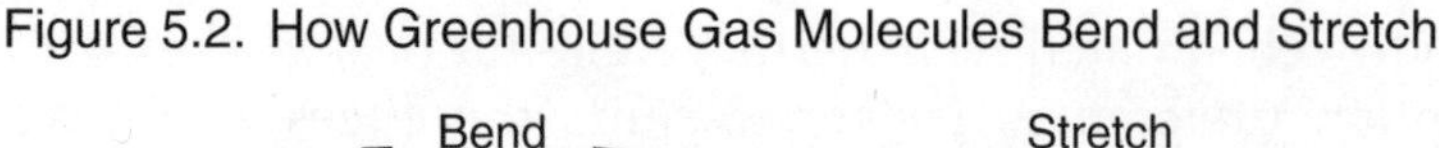

Figure 5.2. How Greenhouse Gas Molecules Bend and Stretch

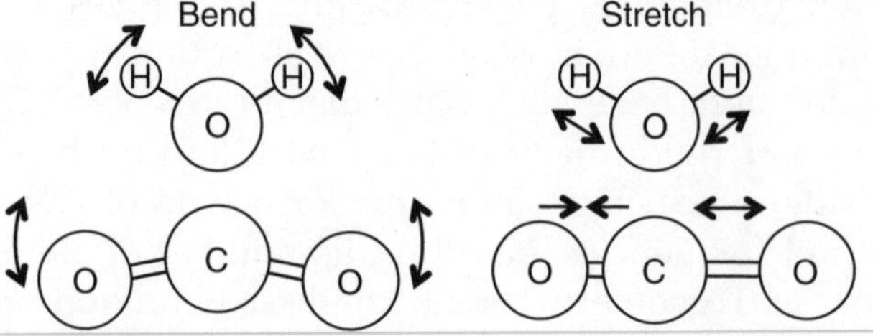

Examples of ways in which two greenhouse gases, H_2O and CO_2, can bend and stretch asymmetrically when they interact with infrared radiation.

closer to the central carbon atom, while the other oxygen atom gets a little farther away. These movements also produce asymmetrical charge distributions. Contrast these dipole-inducing motions to the kind of stretching that a molecule such as N_2 or O_2 can do. Each of these molecules has only two atoms, and the two atoms have the same mass. The atoms can stretch towards or away from one another, but the charges stay symmetrical in this case. N_2 and O_2, the gases that together make up 99 percent of Earth's atmosphere, are not greenhouse gases. Their chemical structure is such that they cannot absorb or re-emit infrared radiation.

The second important factor that determines a good greenhouse gas is its concentration in the atmosphere. Clearly, a gas could have the appropriate chemical structure to be a greenhouse gas, but if its concentration is zero, it does not matter. Add a bit of the gas to the atmosphere, and it will absorb and re-emit some of the radiation at the wavelengths it likes, altering Earth's energy flows, in W/m^2, by some small amount. Increase the concentration, and its influence will increase. Eventually, there will be enough molecules of the gas in the atmosphere, that it will absorb (and re-emit) virtually all the photons coming from Earth's surface that are at its preferred wavelength. It will absorb wavelengths that are close to either side, too – for example, CO_2 interacts with wavelengths at 14.9 and 15.1 microns in addition to those at 15 microns – broadening its absorption bands as its concentration increases, to the limits of its chemical capabilities.

There are two main points to grasp about concentration and effectiveness as a greenhouse gas. The first is intuitive: as concentration increases, the stock of energy in Earth's climate system (in W/m^2) increases, and the greenhouse effect is stronger. The second point is less intuitive: how much stronger is the effect? The relationship between concentration and W/m^2 is logarithmic, rather than linear. Essentially, Earth gains about the same number of W/m^2 every time the concentration of the gas doubles. If CO_2 concentration were to increase from 100 ppm to 200 ppm, Earth's climate system would gain about an additional 4 W/m^2. (This number includes fast feedbacks, such as the response of water vapor – see section 5.2.3.) To gain a *second* additional 4 W/m^2 or so, the concentration must *double* again, to 400 ppm. Imagine, in baseball, if you added a fourth outfielder to the standard three, a team's chances of catching fly balls would increase. But if you already had twenty outfielders, the addition of one more probably would not make much difference – your team would already catch almost all the fly balls. This logarithmic feature means that the effect of adding more of a particular greenhouse gas depends partially on the initial concentration of the gas and on the proportion of its preferred photons it is already intercepting. The higher CO_2 concentrations go, the less influence each additional molecule has on the greenhouse effect.

Partly because of these diminishing returns, adding one more molecule of methane at today's concentrations increases the greenhouse effect more than adding one more molecule of CO_2. But a gas's effectiveness over time also depends on how long it sticks around. For example, carbon monoxide (CO) has a chemical structure such that it can absorb and emit infrared radiation, but it quickly oxidizes to CO_2 in the atmosphere, so its direct influence as a greenhouse gas is insignificant. Similarly, methane, which has a fairly short atmospheric lifetime of about twelve years, also oxidizes to CO_2, the weaker (molecule-for-molecule) but longer-lived greenhouse gas. The CO_2 produced from methane oxidation might stick around for a century or so, and some of it will stay in the atmosphere for thousands of years.

As with any stock-and-flow system, changes in the stock arise from imbalances in flows and depend on how long those imbalances persist. The stocks of all important greenhouse gases have increased in the past century as inflows have exceeded outflows. For each gas, however, its own combination of stock, processes controlling flows, and lifetime influences its long-term role as a greenhouse gas. Water vapor, for example, has the highest stock of any greenhouse gas in the atmosphere, high rates of inflow and outflow, and fast turnover. Because it evaporates and condenses easily on Earth, the stock of water vapor responds quickly to changes in temperature. If, for some reason, Earth's temperature plunged by 10°C tomorrow, the amount of water vapor in the atmosphere would decline dramatically within a couple of weeks.[5] In contrast, CO_2 has a much lower concentration in the atmosphere than H_2O. But there are processes that can *add* CO_2 to the atmosphere very quickly (burning fossil fuels, rapid release from permafrost), while processes that *remove* CO_2 from the atmosphere over the long term are slow (uptake by the deep ocean, rock weathering, burial of organic carbon). In short, the stock of atmospheric CO_2 can increase rapidly, but it decreases only slowly.

The combination of all the factors described above determines which gases are responsible for most of Earth's greenhouse warming. Due to its high concentration and many absorption bands, water vapor is directly responsible for the most W/m^2 change in energy in Earth's climate system: under clear sky conditions (no clouds), it accounts for about 67 percent of greenhouse warming. We will see later, however, that the concentration of water vapor is a feedback, rather than a driver. CO_2 comes in second place, responsible for around 24 percent of total warming, and it is a primary driver on human time scales. All other greenhouse gases make up the remaining 9 percent.[6]

5.1.2. Energy flows in a greenhouse world

Most shortwave radiation from the Sun passes through Earth's atmosphere with little interference; some gets absorbed, but the clear-sky atmosphere is largely transparent to the wavelengths of solar radiation. Earth's surface absorbs solar radiation, heats up, and emits longwave, infrared radiation. Greenhouse gases absorb and re-emit the infrared energy, sending some back towards Earth; they also collide with other gas molecules, warming the air. If there were no gases in the atmosphere able to absorb and re-emit radiation at the wavelengths that Earth emits – if, for example, our atmosphere were 100 percent N_2 and O_2 – this energy would proceed directly back into space and Earth's energy balance would be simple. But greenhouse gases complicate matters.

Imagine the greenhouse gases in Earth's atmosphere as players in a giant, but directionally random, game of catch. Earth emits (throws) a variety of types of "balls" – photons of different wavelengths – away from its surface. If a photon of a particular wavelength happens to encounter a gas capable of absorbing it, the gas will absorb the photon, then re-emit one of the same wavelength. Unlike a game of catch between people, though, the gas has no control over the direction in which it re-emits the photon – it could go anywhere. If it happens to go towards another greenhouse gas capable of absorbing it, the "catch" and "throw" repeat, and a photon goes off in yet another random direction (Figure 5.3). Eventually, some photons happen to get

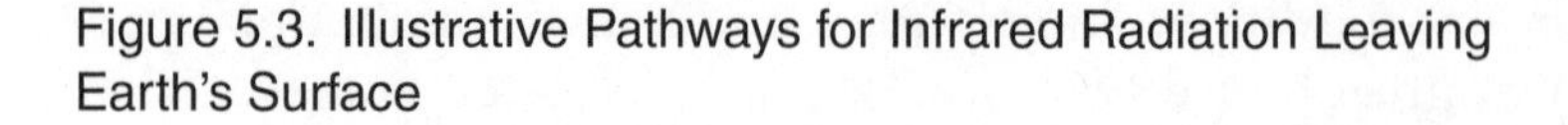

Figure 5.3. Illustrative Pathways for Infrared Radiation Leaving Earth's Surface

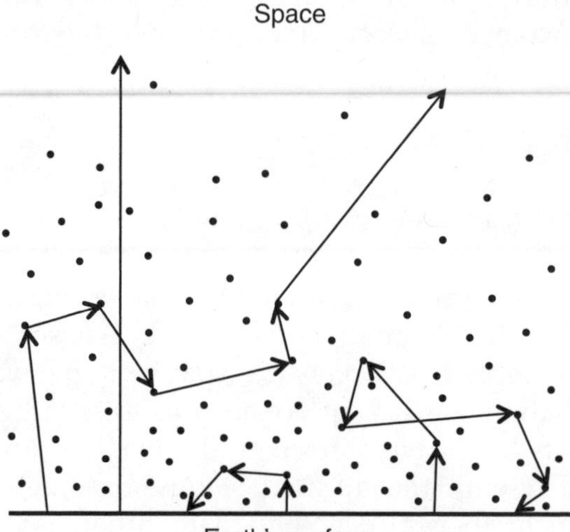

Dots represent greenhouse gases in the atmosphere. Most infrared energy emitted from Earth's surface is absorbed by greenhouse gases, then re-emitted in random directions, slowing the passage of energy into space and warming Earth. Some energy eventually gets re-absorbed by Earth's surface, and some eventually goes out into space. Energy with wavelengths in the atmospheric window (Figure 5.1) proceeds directly to space.

emitted back to Earth's surface, where they are re-absorbed (then re-emitted). Some photons happen to get emitted towards space, and escape the planet without further encounters.

The random directions of emissions are a key part of the greenhouse effect. The photons emitted by Earth's surface always enter from the bottom of the atmosphere. If greenhouse gases always absorbed photons from below, then turned around and emitted them towards space, there would be a constant unidirectional flow of energy from Earth's surface to space, like a human chain passing buckets of water to a fire, and it would have little influence on atmospheric temperature. Now imagine the opposite, that greenhouse gases always returned the energy directly back towards Earth's surface, like a wall returns a ball thrown at it. None of that infrared energy would ever be able to leave the system, quickly leading to catastrophic heating. Instead, the situation is more like a free-for-all water balloon fight. Greenhouse gases emit in random directions and some photons travel back towards the surface, increasing the stock of energy close to the source.

In this game, there are many types of balls in the air. The CO_2 team is throwing 15-micron photons, while the CH_4 team is throwing 7.5-micron photons. In some cases, they have exclusive rights to particular photon wavelengths, and they leave each other's photons alone. In other cases, they interfere with another team's game a bit if both molecules can absorb and emit photons at a particular wavelength. Some types of photons nobody can catch, and the largest wavelength range for these types is about 10 to 12 microns, called the "atmospheric window" (Figure 5.1): there simply are no naturally occurring gases in Earth's atmosphere with the right chemical structure to interact with these photons, so they pass nearly directly from Earth's surface to space without getting involved with the primary greenhouse gases at all.[7]

The net effect of all this absorption and random re-emission is that energy sticks around in Earth's climate system for longer than it otherwise would, making Earth's temperature warmer than it would be without greenhouse gases. In particular,

Earth's surface and lower atmosphere are heated most, because exchanges of energy among gases are most common near the surface, where the gases are most crowded, and less common at higher altitudes, where the air is thinner. Ultimately, though, the amount of energy leaving the top of the atmosphere and returning to space has to balance the energy arriving from the Sun, otherwise Earth's temperature would increase or decrease until these flows came into balance. A similar balance of energy inflow and outflow must occur at Earth's surface, again with heating or cooling as a stabilizing feedback. Notice in Figure 2.6 that the infrared energy emission from Earth's surface is greater (in W/m^2) than the infrared energy leaving the top of the atmosphere. The difference between these two flows is the magnitude of the greenhouse effect.

In short, greenhouse gases slow the passage of infrared energy from Earth's surface back to space. The result is that Earth is warmer than it would be without greenhouse gases. Increases in atmospheric greenhouse gas concentrations raise the planet's equilibrium temperature, while decreases in greenhouse gases lower it.

5.2. The unperturbed carbon cycle and natural greenhouse variability

5.2.1. Carbon stocks and flows

Carbon is nearly everywhere on Earth, in various chemical forms. It is in the bodies of living organisms, in the shells of shell-making organisms, in CO_2 and CH_4 in the atmosphere, in rocks and soils, and dissolved in both fresh and salt water (see Figure 2.5). Most of the carbon – an estimated 50,000,000 gigatonnes (Gton) – is in rocks, present as either organic carbon (bits of dead biological matter) or as $CaCO_3$, the inorganic mineral in the tiny shells that make up limestone rocks. Lots of carbon is dissolved in the oceans, too, most of it (about 38,000 GtonC) in inorganic forms. Of the organic carbon in the oceans (about 600 GtonC), only a little bit (about 3 GtonC) is in currently living organisms. Compared with the carbon reservoirs in rocks and the oceans, the amount of carbon in the atmosphere is small – currently less than 1,000 Gton – but since atmospheric CO_2 controls the greenhouse effect, this stock and the flows in and out of it are most important here.

All the major carbon reservoirs (rocks, oceans, land biota, and soil) exchange carbon with the atmosphere. Annually, the atmosphere swaps hundreds of Gtons of carbon with the terrestrial biosphere and the oceans, and trades just a tiny amount of carbon with rocks and sediments (Figure 2.5). As with any system, it is the balance or imbalance of these flows that determines the stock of carbon in the atmosphere at any particular time.

Some flows – particularly those involving land plants and the surface ocean – happen quickly. During the growing season, photosynthesis takes carbon out of the atmosphere faster than respiration returns it, drawing the atmospheric stock down. These flows reverse during the opposite season, when respiration and decay exceed photosynthesis, and carbon returns to the atmosphere (Figure 5.4). Averaged annually, the natural flows (not considering human influence) associated with photosynthesis and respiration are in balance, with about 120 GtonC traveling back and forth between the atmosphere and land biota each year. In other fast flows, CO_2 molecules constantly cross the boundary between the air and the oceans, dissolving into the water and outgassing back to the atmosphere. This exchange amounts to about 90 GtonC in each direction annually. Other flows – such as rock weathering, which removes CO_2 from the atmosphere, or volcanoes, which return CO_2 from rocks back to the atmosphere (look back again at Figure 2.5) – are just a trickle, with estimates ranging from about 0.05 to –0.2 GtonC/year. Though annually tiny, these flows are important on time scales of millions of years. When flows in and out of the atmosphere are in balance, the atmospheric carbon stock does not change over time. But when they are out of balance, whether for natural or other reasons, the stock of atmospheric CO_2 changes until some stabilizing feedback brings the system back to equilibrium.

Figure 5.4. Atmospheric CO_2 Concentration, 1958–2014

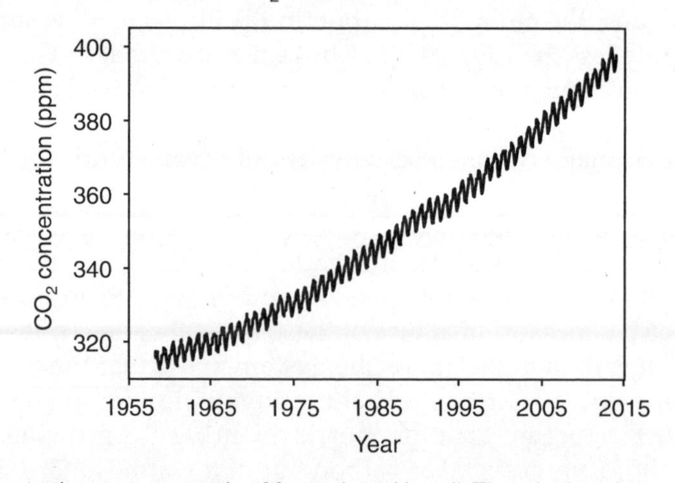

Concentration as measured at Mauna Loa, Hawaii. The wiggles show seasonal changes in CO_2. Atmospheric CO_2 decreases during the growing season (spring to autumn), during which outflow from photosynthesis exceeds inflow from respiration and decay, then increases from autumn to spring, during which inflow exceeds outflow. The upward trend is due to human emissions of carbon into the atmosphere (an extra inflow), which exceed the natural outflows to the land biosphere and oceans.

Sources: United States, Department of Commerce, National Oceanic and Atmospheric Administration, Earth System Research Laboratory, Global Monitoring Division, "Trends in Atmospheric Carbon Dioxide," available online at http://www.esrl.noaa.gov/gmd/ccgg/trends/; and Scripps Institution of Oceanography, Scripps CO_2 Program, "CO_2 Concentration at Mauna Loa Observatory, Hawaii," available online at http://scrippsco2.ucsd.edu/.

5.2.2. Time scales of natural greenhouse variability

Over time scales of months to millions of years, atmospheric carbon dioxide concentration and the magnitude of the greenhouse effect increase and decrease. We have an excellent record of variability in the atmospheric CO_2 stock since the late 1950s, when C.D. Keeling started making direct measurements at Mauna Loa, Hawaii, as Figure 5.4 shows. Farther back in time, we rely on analyses of air bubbles (samples of previous atmospheres) trapped in the Greenland and Antarctic ice sheets. The records from Greenland extend back about 123,000 years[8] and those from Antarctica extend back about 800,000 years.[9] Farther back than the ice core records, reconstructions of atmospheric CO_2 are based on various indirect measurements, such as the density of stomata on fossil leaves (the openings through which plants take in CO_2) and the chemistry of fossil shells and ancient soils.

Without going into details of each measurement, or how the atmospheric CO_2 stock is reconstructed so far into the past, in the next sections we explore records of atmospheric CO_2, the explanations of these patterns over time, and what the patterns tell us about how Earth's climate and the carbon cycle work. Better understanding of how atmospheric CO_2 and climate evolved in the past can help us understand the implications of current and future greenhouse gas emissions.

5.2.2.1. The long-term view: Hundreds of millions of years

Over the very long term, atmospheric carbon dioxide concentrations go up and down in response to slow-acting processes, most of which are related to plate tectonics in some way. The primary processes that remove CO_2 from the atmosphere on long time scales are chemical weathering of silicate rocks and burial of organic carbon (see section 2.2.1.4.). The primary processes that add CO_2 to the atmosphere on long time scales are volcanic eruptions and, to a lesser extent, weathering of carbon-rich rocks such as coal.

We will start by exploring an important long-term stabilizing feedback involving the Sun, greenhouse gases, chemical weathering, and volcanoes. As we saw in Chapter 3, the Sun's output is increasing over time. As it does, Earth receives more solar radiation and its temperature rises. At higher temperatures, for a given atmospheric CO_2 concentration, chemical weathering rates increase, since many chemical reactions happen faster at higher temperatures. The increased chemical weathering draws CO_2 out of the atmosphere, decreasing the greenhouse effect and cooling the planet. At new, lower temperatures, chemical weathering rates slow down to the point where outflow from weathering matches inflow from volcanoes. Atmospheric CO_2 is thus stabilized, at a lower concentration than previously, compensating for the higher solar radiation. Over the very long term, the greenhouse effect decreases as solar output increases, generally stabilizing the planet's temperature.

But Earth hasn't been a constant temperature over its lifetime. Large perturbations, involving plate tectonics, biology, and sometimes the two in combination, overlay this long-term stabilizing feedback. A major cold period about 2.5 billion years ago likely resulted from the evolution of oxygen-producing photosynthetic organisms. When they started pumping oxygen into a previously oxygen-free atmosphere, the newly released oxygen oxidized everything it could, including atmospheric methane (CH_4), which converted to CO_2. Since CO_2 is a weaker greenhouse gas, molecule-for-molecule, than methane, the greenhouse effect decreased in magnitude and Earth cooled. A cold period about 300 million years ago likely can be explained by the formation of the supercontinent Pangaea and extensive land area covered in swamps. The process of tectonic plates coming together formed mountains and promoted weathering, which drew CO_2 out of the atmosphere. In addition, the swamp plants, like any other photosynthesizers, took CO_2 out of the atmosphere as they grew. But when the plants died, lots of their organic carbon got buried in those swamps instead of quickly returning the carbon to the atmosphere. Mountain building plus extensive swamps decreased the greenhouse effect and cooled Earth.

Two other cold periods likely owe their low temperatures to excess chemical weathering. Geologic evidence suggests that, about 650 million years ago, Earth was a cold "snowball," partly because the continents happened to be clustered in the tropics. Lots of rock surface area in a regionally warm, wet climate resulted in lots of chemical weathering, which drew CO_2 out of the atmosphere and decreased the greenhouse effect. The poles became ice covered, and the ice expanded towards the equator as the planet cooled. Eventually, the amplifying ice-albedo feedback (Chapter 4) helped the ice expand all the way into the tropics. And in the past 50 million years, Earth has been cooling gradually as rock weathering associated with the formation of the Himalayas and the Tibetan Plateau – a massive mountain-building event from the collision of the Indian and Eurasian plates – decreases atmospheric CO_2.

After each cold period, the greenhouse gradually recovered as volcanoes spewed CO_2 over time and inflows and outflows moved back towards balance. Volcanism itself, associated with faster or slower tectonic plate motions, can also perturb the climate system. For example, the dinosaurs lived in a warm period associated with increased volcanism on the sea floor, which increased atmospheric CO_2 and promoted warming.

The greenhouse effect has strengthened and weakened on long time scales over Earth's history as tectonic plates move around, mountains are built and erode, volcanoes are more or less active, and conditions for plants change. If viewed within the enormous sweep of geologic time, climate conditions on Earth today are relatively cool (Figure 3.5).

5.2.2.2. The medium-term view: Hundreds of thousands of years

So, if our context is the entirety of Earth's history, we live in a "cold" period. But zoom in on the past million years or so and in that time frame today's climate is relatively warm with relatively high CO_2 concentrations (Figure 3.6). It is these glacial-interglacial cycles that comprise the "background" climate setting on Earth today (see section 3.2.2. about drivers of these climate cycles). Deviations from this glacial-interglacial context are more relevant to humans than variability on time scales of tens to hundreds of millions of years.

We know, from ice core measurements, that atmospheric CO_2 was lower during glacial periods and higher during interglacial periods. What carbon cycle processes act on time scales relevant for glacial-interglacial cycles? How did atmospheric CO_2 yo-yo up and down over the past million years or so? Based on evidence from the carbon chemistry of shells buried in ocean sediments, water in the deep oceans contained more carbon during glacial periods than during interglacial periods. The evidence suggests that carbon stored in biomass on land gets transferred to the deep oceans during cold glacial periods, then back again during warm interglacial periods.[10] What is going on? Certainly, the massive ice sheets that grew on land in the northern hemisphere would have displaced large tracts of what is today boreal forest, and the carbon from those trees went somewhere. Evidence related to wind-blown dust indicates that the glacial world was drier, with more expansive deserts, which are biomass-poor ecosystems. So, a variety of evidence indicates less carbon was stored in the land biomass reservoir during glacial periods.

If this carbon went into the deep oceans for temporary storage, how did it get there? Two primary pathways, one physical and one biological, are the likely candidates. The physical pathway has to do with carbon exchange between the oceans and the atmosphere. The biological pathway involves pieces of dead organic matter sinking from the surface of the oceans downward.

When you open a can of warm soda, a bunch of gas comes spurting out, but if the soda is cold, there is just a gentle fizz. This exemplifies the behavior of CO_2 as it dissolves into and degasses from the oceans. In colder water, more CO_2 dissolves and stays in solution; in warmer water, excess CO_2 outgasses back into the atmosphere. Using ice core data, scientists have studied the relationship in time between atmospheric CO_2 and temperatures (Figure 3.6). Imagine, for example, that a change in Earth's orbit helps melt an ice sheet a little, lowers albedo, and warms the planet. A warmer ocean would outgas a bit of the CO_2 it formerly held. That CO_2 would add to the greenhouse effect, warming the planet further, promoting more outgassing, and so on, in an amplifying feedback. The same feedback works in reverse during a period of cooling. Thus, on ice-age time scales, CO_2 is involved in an important feedback with temperature.

In the oceans, organisms are concentrated near the surface, where photosynthetic phytoplankton live. Carbon and nutrients largely get recycled and reused near the surface, but there is some leakage of organic material downward, as fecal pellets and dead organisms sink through the water column. During the ice ages, drier, windier conditions might have enhanced transport of nutrients from land – particularly micronutrients such as iron, which limits phytoplankton growth in some areas. With an inflow of nutrients, biological productivity likely increased, which, in turn, increased the leakage of organic carbon downward to temporary storage in the deep oceans. There was, of course, a return flow in the form of the upwelling of deeper, carbon- and nutrient-rich water back to the surface. As in any stock-and-flow system, if carbon flow downward equals flow upward, there is no change in stock. But if an imbalance persisted for some time – that is, if carbon flow downward exceeded flow upward – then the stock of carbon in the deep oceans would increase. Various lines of observational and modeling evidence indicate that the glacial deep oceans did not circulate as quickly as they do today, and that sluggishness would have kept the carbon down there for longer. The general time scales of deep ocean overturning are in the range of hundreds to thousands of years, which reasonably aligns with the time scales of glacial-interglacial transitions. One would not look, for example, to mountain building or to seasonal cycles in photosynthesis to explain perturbations in carbon stocks on glacial-interglacial time scales. The former is much too long and the latter much too short to matter.

So, the combination of processes that move carbon among the atmosphere, biosphere, and deep oceans have resulted, over the past million years or so, in atmospheric CO_2 concentrations cycling back and forth between about 180 ppm during cold glacial periods and about 280 ppm during warm interglacial periods. That is a change of about 210 GtonC (100 ppm CO_2) in the stock of carbon in the atmosphere.

5.2.2.3. *Abrupt change: Analogue for our future?*

We sometimes assume that the greenhouse effect and Earth's climate change rather slowly and smoothly. This mental model of climate change is somewhat comforting: if climate changes slowly and smoothly, then humans should have time to respond. But what if climate changes take place abruptly? Can natural and human systems address or adapt to such changes?

Examples from the past can give us insights into the kinds of havoc abrupt climate changes can wreak.

A classic example happened about 55 million years ago, when a dramatic event perturbed the carbon cycle and Earth's climate.[11] Something caused a rapid transfer of carbon to the atmosphere from some other reservoir. Atmospheric CO_2 increased, global temperatures warmed by 5°C or more, the oceans turned acidic, and many marine organisms went extinct – in short, a planetary disaster. Scientists are still sorting out the details: where did the carbon come from? how much carbon went into the atmosphere? Answering these questions involves aligning estimates of the degree of ocean acidification, the magnitude of global temperature change, and the chemistry of potential carbon sources, from which we can learn useful information about how fast such an event unfolds and how long Earth takes to recover from rapid and large-scale perturbations of the carbon cycle.

Estimates of how fast the carbon release happened 55 million years ago range from a few thousand to about 20,000 years.[12] That is fast in geologic terms, slow in human terms. In comparison, the available evidence indicates that today's rates of carbon inflow to the atmosphere and associated ocean acidification are happening ten to thirty times faster than occurred 55 million years ago.[13] And recovery took a long time. Eventually, atmospheric carbon dioxide decreased again. Eventually, the oceans recovered their previous pH and marine organisms evolved and carried on. But it took more than 100,000 years for carbonate sediments in the oceans to neutralize the acid that had been introduced, and even longer for atmospheric CO_2 to decrease to the previous background concentration, thanks to the slow processes of weathering silicate rocks on land.[14] At today's rate of change, there is even less time for natural flows of carbon between the different reservoirs to respond and balance out. This 55-million-year-old analogy tells us that rapid changes to the carbon cycle such as those we see today can have surprisingly deleterious consequences. Recovery is likely to be slow, unless we begin deliberately increasing the outflow of carbon from the atmosphere to balance inflows.

5.2.3. Feedbacks involving the greenhouse effect

Adding greenhouse gases to or removing them from the atmosphere not only directly alters greenhouse warming, but also perturbs various feedbacks – some stabilizing, others amplifying. The most important stabilizing feedback is the longwave radiation feedback, which involves energy radiated from Earth's surface and atmosphere. The most important amplifying feedback is the water vapor feedback, in which water vapor, a greenhouse gas, amplifies temperature changes. Two other important feedbacks are the lapse-rate feedback (stabilizing), which involves changes in how steeply atmospheric temperature drops with increasing altitude, and the longwave cloud feedback (amplifying), but many other feedbacks in the climate system also respond to perturbations in energy flows.

Currently, due to recent increases in the greenhouse effect, Earth is absorbing, on average, more energy than it emits, with the imbalance amounting to between +0.5 and +1 W/m^2.[15] This excess inflow of energy warms Earth and sets in motion the most important stabilizing feedback in the system: the longwave radiation feedback. As Earth warms, its temperature increases, so that the energy it radiates (the energy outflow) also increases (see equation 2 in section 3.1.1.), counteracting the warming (Figure 5.5, panel A). Eventually, energy outflow increases until outflow again matches inflow, and Earth stabilizes at a new (in this case, higher) temperature. The longwave radiation feedback thus helps keep Earth's temperature changes in check.

To erase the current imbalance, Earth needs to warm by about 0.6°C or so,[16] which is often termed the warming "in the pipeline." This means that, if we could freeze greenhouse gas concentrations at today's values, we would still see at least that much additional warming simply due to Earth's system catching up with the energy flow imbalance. But if we could reduce the energy imbalance by, for example, decreasing the concentration of greenhouse gases in the atmosphere, there would be less catching up to do, and the longwave radiation feedback would stabilize Earth's temperature at a lower value.

The longwave radiation feedback is crucial for climate stability, but other feedbacks reinforce warming. In the amplifying water vapor feedback, warming causes an increase in water vapor (a greenhouse gas) in the atmosphere, which strengthens the greenhouse effect (Figure 5.5, panel B). The reverse is also true: cooling decreases atmospheric water vapor, which decreases the greenhouse effect, promoting more cooling. This is a fast-acting feedback, since water changes phase quickly. Where water is available, its molecules constantly exchange between liquid and gaseous states (and solid state, if temperatures are cold enough). If temperatures are warm, water molecules move around rapidly and more of the available water will be in the gas phase; at cold temperatures, more of the

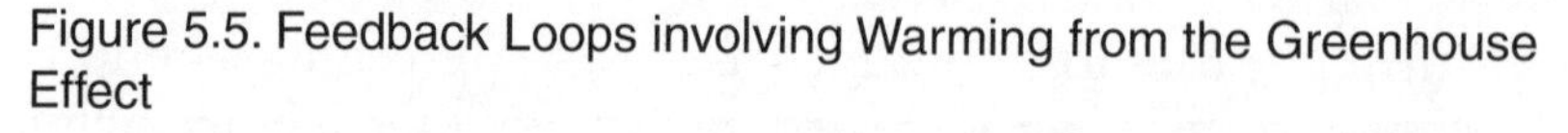
Figure 5.5. Feedback Loops involving Warming from the Greenhouse Effect

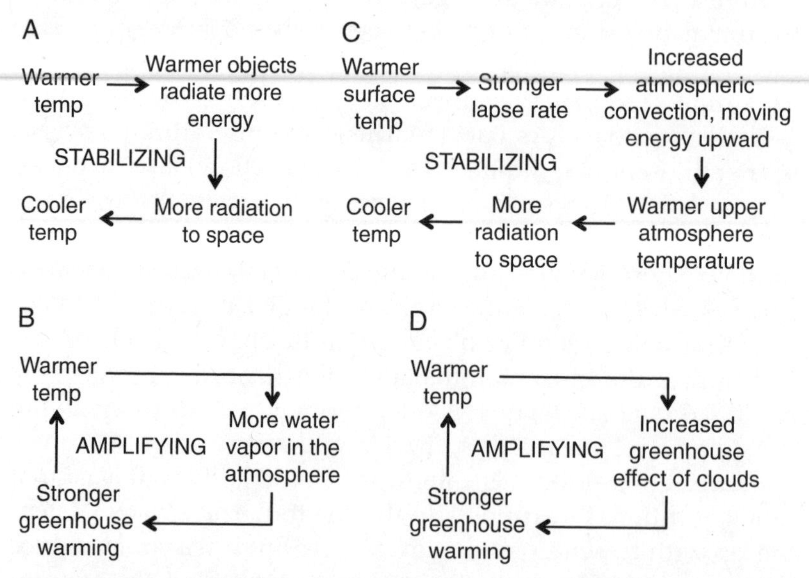

Panel A shows longwave radiation feedback in a warming world (stabilizing). Panel B shows water vapor feedback in a warming world (amplifying). Panel C shows lapse-rate feedback in a warming world (stabilizing). Panel D shows cloud feedback in a warming world (amplifying).

available water will be in liquid or solid phases. A bucket of water in a sauna quickly evaporates; a glass of ice water in a warm room quickly collects liquid condensation on the outside of the glass. In the atmosphere, more liquid water evaporates when it is warm, and water vapor condenses back into water droplets when it cools. Those droplets might fall as rain, or they might evaporate to vapor again.

Because water vapor responds so quickly to temperature changes, it cannot drive global temperatures directly; instead, it is relegated to a supporting role, though one with a big impact. We can put more water vapor into the atmosphere temporarily by boiling pots of water on the stove, by running sprinklers, or by releasing water from airplanes, but the water would stick around in the atmosphere only if the temperature is right. Instead of changing water vapor directly, however, we are changing the temperature by altering the atmospheric concentration of long-lived CO_2. Warming from CO_2 sets the water vapor feedback in motion, which effectively doubles the heating impact of CO_2 alone. CO_2 is the ringleader; water vapor, a loyal, dependable follower.

Water vapor is a player in an important stabilizing feedback, too: the lapse-rate feedback. The lapse rate describes how atmospheric temperature gets colder with increasing altitude. A typical lapse rate is something like 7°C per km, so that, at 10 km altitude, the air is about 70°C colder than at the surface. The stronger the greenhouse effect, the greater the difference between the energy leaving Earth's surface (in W/m^2) and the energy emitted at the top of the atmosphere, the stronger the temperature contrast between the warm lower and cold upper atmosphere, and the steeper the lapse rate.

Consider the upper atmosphere, which, since it is colder, emits less radiation than the warmer lower atmosphere. The upper atmosphere radiates in all directions, including into space, and receives energy from below. If inflow from below exceeds radiation into space, the air will warm, and the lapse rate will become less steep, until inflow and outflow balance again. If the upper atmosphere gets too warm, radiation into space will exceed inflow from below, and the air will cool again, steepening the lapse rate, and eventually balancing the energy flows.

So how does water vapor fit in? When water evaporates from Earth's surface, the lighter H_2O molecules displace heavier N_2 and O_2 molecules in the atmosphere, decreasing the overall density of the air close to the surface. If this air near the surface is less dense than the air above it, the warm, humid air will rise from the surface. As it rises, the air expands and cools, and the water vapor condenses again, releasing latent heat to the atmosphere, but now at a higher altitude. This process, called convection, helps transfer energy upward in the atmosphere, and helps decrease the lapse rate. The now-warmer upper atmosphere can radiate more energy into space than before, which a stabilizing feedback in response to surface warming (Figure 5.5, panel C).

And last, clouds respond to changing temperature and can amplify temperature changes through their interactions with longwave radiation. Clouds are tricky, because they both reflect incoming solar radiation (Chapter 4) and contribute to the greenhouse effect. Different types of clouds differ in their relative contributions to albedo (cooling) versus greenhouse (warming). Cold, thin clouds at high altitude contribute more to greenhouse warming than to albedo; therefore, increasing high clouds yields net warming. Clouds at low altitude are thicker and more highly reflective, but some of these clouds might evaporate in a warming world. The currently available evidence suggests that the greenhouse effect from clouds likely increases in a warming world (Figure 5.5, panel D), and that the sum of cloud feedbacks, including both albedo and greenhouse effects, is likely to be amplifying overall.[17]

If we add up the effects of all the feedbacks in the climate system, the result, fortunately, is stabilizing. If the sum of feedbacks were amplifying, we would be headed towards a very hot or very cold world. In aggregate, feedbacks nudge the planet towards temperature stability, even though the value of that stable temperature varies with time.

5.3. Anthropogenic interference

5.3.1. Perturbed stocks, flows, and chemical fingerprints

An increasingly strong greenhouse effect is the main culprit in the global heating of the past one hundred years or so. But how do we know that humans are responsible? To solve a crime, investigators might use several lines of evidence – the actions of suspects, fingerprints, DNA – to identify who was involved. If the collection of evidence coalesces into a coherent picture, the investigators have increased confidence that they have correctly reconstructed the scene. Similarly, various lines of evidence point to humans as the party responsible for increasing atmospheric CO_2 concentrations and, thus, for greenhouse heating.

Humans perturb the carbon cycle by increasing flows of carbon to the atmosphere. Fossil fuel burning and land-use change are the primary human actions by which this is happening. Cement making, from limestone, also transfers carbon to the atmosphere. Here, we examine three primary lines of evidence linking human activities to changing atmospheric CO_2: (1) stock-and-flow data, which show that human emissions align well with the increase in atmospheric stock; (2) chemical evidence using carbon isotopes, which allows us to identify the source of the new atmospheric carbon; and (3) chemical evidence that shows an expected decrease in atmospheric oxygen.

First, if we can reconstruct human activities over time, we can estimate the effect of those activities on atmospheric CO_2. How much fossil fuel, of what types, containing how much carbon, has been burned each year since the start of the Industrial Revolution? The Carbon Dioxide Information Analysis Center compiles data on emissions from fossil fuels and cement production back to 1751, based on human historical records of extraction and trade.[18] Those data show that carbon emissions from those sources have grown, roughly exponentially, from about 0.003 GtonC/year in 1751 to about 9.5 GtonC/year in 2011. Land-use change is currently adding about another 0.7 to 1.5 GtonC/

year.[19] These are the primary human-related inflows of carbon to the atmosphere today.

The industrial age's burning of fossil fuels gets most of the credit for human-induced greenhouse warming, but starting about 8,000 years ago, atmospheric CO_2 concentrations began to increase in a pattern unlike in any similar interglacial period in the past 800,000 years.[20] What was different this time? One compelling explanation is that humans began clearing forested land for agriculture, which decreased carbon stored in terrestrial plants and increased CO_2 in the atmosphere. Similarly, about 5,000 years ago, atmospheric concentrations of methane began an anomalous rise, a change that coincides with the expansion of land use for rice cultivation, which floods land and creates swampy conditions ideal for the production of CH_4.[21] Humans have been perturbing carbon stocks for a long time.

So, atmospheric CO_2 has increased over the past 8000 years, and the pace of the increase has greatly quickened since the Industrial Revolution. Interestingly, however, the rise is not as fast as it would have been if *all* carbon emissions by humans had stayed in the atmosphere; instead, every year, only about 45 percent of our emissions stay in the atmosphere adding to the stock of atmospheric CO_2, while the other 55 percent or so is absorbed by the oceans and terrestrial ecosystems.

Documented human activities seem to align well with measured changes in the stock of atmospheric carbon. But how confident are we that it is human actions, not some other source of carbon, that are causing the increase? To answer this question, we can examine the particulars of the chemistry of atmospheric carbon.

Carbon isotopes allow us chemically to "fingerprint" carbon sources. Here is how it works. The chemical element carbon has six protons, but it comes in slightly different varieties. The vast majority (about 99 percent) of carbon atoms also have six neutrons, which, with the six protons, add to an atomic mass of 12. That common, garden-variety carbon is carbon-12, written as ^{12}C. Some carbon atoms (about 1 percent) have an extra neutron, yielding ^{13}C. This form of carbon is still stable – it is just a slightly heavier isotope. And there is carbon-14 (^{14}C), a naturally occurring radioactive isotope of carbon with two extra neutrons, present in just trace amounts. These three isotopes, ^{12}C, ^{13}C, and ^{14}C, help us determine the source of the carbon that is being added to the atmosphere, because different sources contain different relative proportions of these isotopes.

Photosynthesis by plants is a key process that helps untangle the problem of source. Plants take CO_2 out of the atmosphere, and every CO_2 molecule has ^{12}C, ^{13}C, or ^{14}C as its carbon atom. But, for reasons of energy expenditure, plants do not take in the different carbon isotopes in the same proportion as they are available in the air. Instead, like a cashew fan picking through a bowl of mixed nuts, plants preferentially use ^{12}C over ^{13}C. So plant matter ends up with a smaller ("lighter") ratio of ^{13}C to ^{12}C, while the atmosphere ends up a little "heavier," since the ^{13}C mostly gets left behind, like the peanuts in the bowl. The ancient plants from which fossil fuels formed also preferred ^{12}C over ^{13}C when they were photosynthesizing, so coal, oil, and natural gas all have "light" carbon isotopic compositions. When plants or fossil fuels rot or burn, the carbon in them oxidizes to CO_2, and their "light" carbon returns to the atmosphere. Deforestation (if there is net biomass loss) and burning fossil carbon thus both help make atmospheric CO_2 isotopically "lighter," on average. And, in fact, measurements show that the ratio of ^{13}C to ^{12}C in atmospheric CO_2 has decreased – become "lighter" – over time.[22] These findings, combined with changes in stock, allow us to eliminate other candidate carbon sources. Volcanoes, for example, emit CO_2 with isotopic proportions much closer to atmospheric proportions. They do not emit enough "light" carbon to account for the change in the chemical fingerprint we see – in fact, they do not emit much carbon at all compared with the amount humans do. Carbon isotope evidence thus points to human activities as the carbon source.

How does carbon-14 help? This radioactive isotope has a relatively short half-life, compared with the age of fossil fuels: ^{14}C that was incorporated into a plant during its life radioactively decays away virtually to nothing in about 50,000 years.

But much of the fossil fuel we burn formed much earlier than that – for example, lots of coal formed around 300 million years ago – so that essentially all the ^{14}C that originally might have been present in these fossil fuels decayed long ago, giving fossil fuels a very "light" ratio of ^{14}C to ^{12}C. And we have measured that the ratio of ^{14}C to ^{12}C in the atmosphere has decreased over the time humans have been burning fossil fuels.[23]

Plant growth and decay are also chemically linked to flows of oxygen to and from the atmosphere. Photosynthesis adds O_2 to the atmosphere, and respiration, decay, and the burning of plant matter take O_2 out of the atmosphere as carbon from the plant matter recombines with oxygen, forming CO_2. On an annual basis, if these flows of oxygen in and out of the atmosphere match, O_2 stays constant. But burning fossil fuels or plant biomass for net deforestation increases the outflow of O_2 from the atmosphere. O_2 concentrations have measurably declined over time due to these activities.[24]

In summary, as CO_2 concentrations in the atmosphere have gone up, the isotopic composition of atmospheric CO_2 has become lighter, and atmospheric O_2 has decreased, as expected, since we know how fossil fuel burning and deforestation affect these parameters. There is no plausible source of carbon, other than human activities, to explain these combined trends.

5.3.2. Cumulative carbon emissions: A budget

As we add CO_2 to the atmosphere, we change energy flows in Earth's climate system, and maintain or exacerbate energy imbalances. How much change do we think is too much? For mitigation and adaptation planning and action, we want to know the likely effects of our perturbations. Consider temperature. Internationally, people have agreed that limiting the rise in Earth's global average temperature to less than 2°C above pre-industrial values is desirable – though some would argue even that limit is too high. What do we have to do to meet that target?

One of the most useful, straightforward metrics relates temperature linearly to cumulative carbon emissions: the sum of all carbon emissions since humans began emitting carbon into the atmosphere.[25] The more total carbon we add to the atmosphere, the higher the eventual "peak" global temperature, similar to how the peak water height in a reservoir behind a dam depends on cumulative precipitation in a river basin, unless outflow is manipulated.

The approach used to define this cumulative carbon-temperature relationship takes into account various carbon cycle responses to increasing atmospheric CO_2. For example, in section 5.1.1., we discussed how adding a molecule of CO_2 to the atmosphere yields diminishing returns for the overall greenhouse effect as CO_2 concentrations increase. But the approximately 45 percent of emissions that stays in the atmosphere each year (section 5.3.1.) is expected to increase as CO_2 concentrations increase and natural uptake pathways (such as the oceans and land biomass) become saturated. It is likely that, in the future, with rising atmospheric CO_2 concentrations, a greater fraction of emissions will stay in the atmosphere for longer. These two effects – diminishing returns per molecule and an increased proportion sticking around in the atmosphere – essentially cancel one another out and simplify the relationship between cumulative CO_2 emissions and temperature. Current estimates are that every 1,000 Gton of cumulative carbon emissions yields about 1.8°C warming above pre-industrial temperatures.[26]

This number allows us to define a carbon budget appropriate for a target temperature. Suppose we set a total emissions budget of about 1,100 GtonC, which would be about equivalent to a 2°C temperature rise. We have already used up about half our allowance, leaving us room to emit an additional, say, 550 GtonC, if a 2°C temperature rise remains our target. At about 10 GtonC/year (as of 2011) and rising, staying within budget is a formidable challenge, but possible. Suppose we meet that target for cumulative carbon emissions. Even if we do, a 2°C warmer world would look different in terms of sea level, the hydrologic cycle, the frequency of extreme weather events, and other features, which we discuss in Chapter 8. In addition, amplifying carbon cycle feedbacks that release carbon currently stored in permafrost might be set in motion, and ice sheet disintegration might accelerate. Potentially large adaptation efforts might be required.

We humans are a powerful force on Earth. We are perturbing the carbon cycle and enhancing the greenhouse effect. For now, the net result of our activities is to increase inflows of carbon to the atmosphere. Perhaps, in the future, we will be able to increase outflows at a significant magnitude and restore some sort of balance. Hundreds of millions of years ago, single-celled organisms managed to change Earth's atmospheric composition; it is unsurprising that a creative and resourceful species like ourselves has done the same. As we begin to grasp the magnitude of these changes, however, we are also pressed to come up with transformative solutions to this problem of excessive carbon inflow: the central topic in Chapter 6.

CHAPTER SIX

Climate Change Mitigation: Reducing Greenhouse Gas Emissions and Transforming the Energy System

6.1. Reducing greenhouse gas emissions: An overview
6.2. The global energy system
6.3. Mitigation strategies

6.3.1. Demand-side mitigation: Energy efficiency and conservation

6.3.1.1. Energy-efficient technologies
6.3.1.2. Conservation and behavior change

6.3.2. Supply-side mitigation

6.3.2.1. Wind power
6.3.2.2. Solar power
6.3.2.3. Biomass and biofuels
6.3.2.4. Geothermal energy
6.3.2.5. Tidal power

6.3.3. Carbon capture and storage

6.3.3.1. Carbon capture and storage
6.3.3.2. Carbon sequestration

6.4. Fostering accelerated and transformative mitigation

MAIN POINTS:

- Climate change mitigation means tackling the causes of climate change, rather than the consequences. This involves either reducing our emissions of greenhouse gases or enhancing carbon sinks.
- Demand-side mitigation involves reducing the amount of energy we require, through either conservation or energy efficiency.
- Supply-side mitigation involves transitioning to renewable, low-carbon sources of energy. We already have a suite of renewable and sustainable technologies at our disposal, each with its particular benefits and costs, but also its technical challenges.
- Carbon capture and storage is the process of stripping CO_2 from fossil fuels as we process or consume them, and storing it in geological formations or the oceans.

A race is under way. From Iran to Australia, microscopic green plants are growing, multiplying, with one fate in mind: to fuel the thousands of airplanes that cut across the sky every day. The potential rewards are spectacular, as consumers, regulators, and companies increasingly become aware of the carbon burden air travel imposes. Indeed, a dramatic shift in the emissions landscape might be closer than we think. The German airline Lufthansa, for instance, recently partnered with an Australian biotechnology company to create algae-based jet fuel on a massive scale. This is just one pocket of innovation to combat climate change, and many more are springing up around the globe.

Technological innovations such as algal fuel, however, do not emerge in isolation, but arise out of, and are linked to, economic and social development, ecological integrity, and human vulnerability. A strong theme throughout this book is the myriad linkages between human and Earth systems. As Figure 6.1 shows, however, the depth and complexity of these linkages are such that responding to climate change is not as simple as coming up with a new technology. In this chapter, we focus on the interactions that occur on the left side of Figure 6.1, particularly with respect to mitigation. In Chapter 9, we will shift our focus to the right side of this schematic.

Mitigation is another term for getting at the "roots" of the climate change problem. The goal is to prevent further climate change before it occurs, rather than deal with the impacts of climate change once it has begun. This is a slightly different meaning for the word than the way it might be used in common conversation, where mitigation often means cleaning something up after it has happened. Mitigation has been the most commonly proposed policy response to climate change since the evidence for human interference with the planet's delicate climatic balance began to emerge. The reason for this is part politics and part optimism. Until recently, it has been politically unpalatable to suggest that we are unable to address the climate change challenge, so some groups view adaptation (or responding to the impacts of climate change) as an admission that we will fail to solve the problem of the human contribution

Figure 6.1. Linkages among Human and Earth Systems, Climate Change, Adaptation, and Mitigation

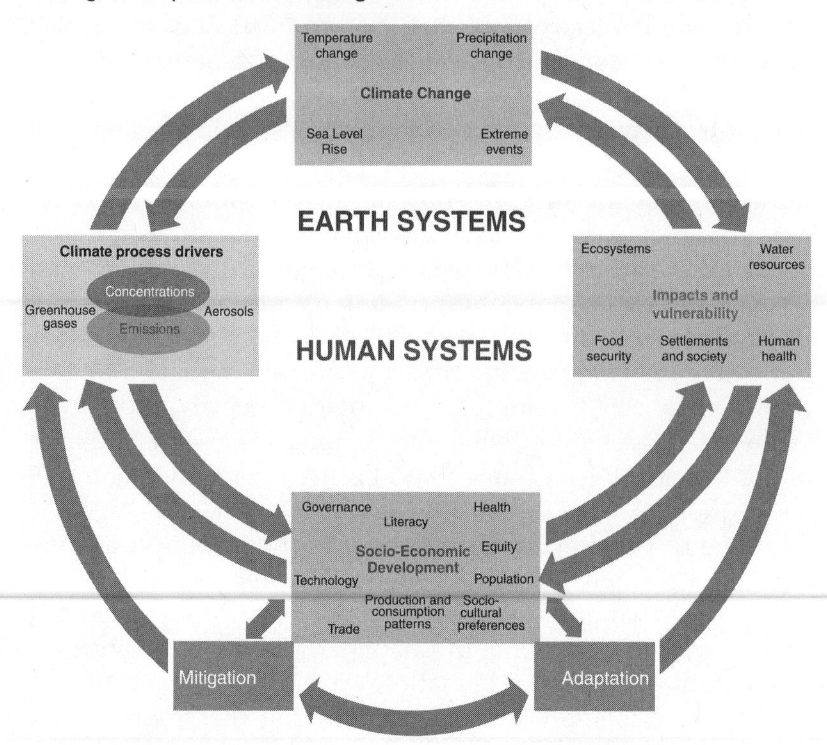

Source: Intergovernmental Panel on Climate Change, *Climate Change 2007, Synthesis Report: Contribution of Working Groups I, II and III to the Fourth Assessment Report of the Intergovernmental Panel on Climate Change*, ed. R.K. Pachauri and A. Reisinger (Geneva: IPCC, 2008), figure 1.1.

to climate change. Others have suggested that a conversation about adaptation requires some acknowledgment that climate change is actually occurring – whether caused by humans or not – whereas mitigation could be supported under the banner of energy independence or urban livability. As the evidence for climate change builds and the human and ecological toll increases, the case for both widespread adaptation and accelerated mitigation has solidified.

Since humans are contributing to accelerating climate change through the emission of massive quantities of greenhouse gases, mitigation involves either reducing our emissions of greenhouse gases into the atmosphere or enhancing the ability of Earth, including its oceans and forests, to absorb carbon. Accordingly, we need to address the fundamental structure of our energy system and look at how we can reduce our energy consumption and make the transition to renewable energy. We also need to explore carbon sequestration and carbon capture and storage. By the end of the chapter, you should have an understanding of the most compelling solutions at our disposal as humans begin to address the challenges of climate change.

6.1. Reducing greenhouse gas emissions: An overview

When we speak of greenhouse gas emissions, we are referring to a flow of greenhouse gases into the atmosphere. Fossil fuel burning and the decomposition of organic matter are the primary sources of greenhouse gas emissions, and result from transportation, industrial processes, residential and commercial heating and cooling, waste treatment and disposal, forestry, and agricultural practices. Greenhouse gas concentrations, in contrast, refer to the quantity of greenhouse gases in the atmosphere as a result of the emissions that have taken place.

In the quest to reduce greenhouse gas emissions, one often hears of "targets" to be achieved. This should be a relatively simple affair to understand, but the numbers we see concerning "targets" on the news or heralded by decision-makers often mask crucial differences in ambition. An important distinction when it comes to reducing greenhouse gas emissions, for instance, is the difference between "intensity-based" and "absolute" greenhouse-gas-reduction targets. The distinction will be important later in the book, when we discuss various policy tools that are at our disposal to mitigate climate change.

Intensity-based targets refer to reducing the amount of greenhouse gases we produce per unit of something. For instance, we might agree to produce fewer greenhouse gases per unit of gross domestic product (GDP), or per capita, or per barrel of oil. By this measure, however, the total amount of greenhouse gases we emit can go up if the size of our economy grows, or the population increases, or the total production of oil goes up. For instance, the Canadian province of Alberta requires companies that emit more than 100,000 tonnes of greenhouse gases per year to reduce those emissions by 12 percent per unit of production each year after 2007.[1] Companies can reach this target in a number of ways. First, they can make performance improvements. Second, they can purchase carbon offsets – emissions reductions to compensate for, or offset, the emissions of green-house gases elsewhere. In this scenario, the emitter typically pays a third party to implement an emissions-reduction project, the benefits of which accrue to the emitter – and pay $15/tonne over the target into a provincially managed fund. And third, they can purchase emissions credits. Despite these policies, total greenhouse gas emissions in Alberta (in CO_2 equivalent) increased from 165.6 megatonnes in 1990 to 233.3 megatonnes in 2010.[2] Alberta is certainly not alone in using intensity targets. In 2002, for instance, the United States set a voluntary goal of reducing the greenhouse gas intensity of the economy by 18 percent between 2002 and 2012. US emissions intensity has declined, but despite the global recession that began in 2008, total emissions were around 8 percent higher in 2013 than they were in 1990.[3]

Absolute greenhouse-gas-reduction targets, in contrast, refer to reducing the total amount of emissions humans produce, regardless of any other factor such as a growing population or a growing economy. This could be a commitment to reducing the *total* amount of greenhouse gases that we emit (the flow) or that are in the atmosphere (the stock). An example would be the commitment Canada made in the Kyoto Protocol to reduce its total annual emissions to 6 percent below 1990 levels by the 2008–2012 period. In 2009, the United States committed to reducing greenhouse gas emissions by 17 percent below 2005 levels by 2020 (an absolute reduction target).[4] A further example is the suggestion that we should aim to limit the concentration of greenhouse gases in the atmosphere to 350 ppm.

Figure 6.2 illustrates the controversy behind intensity-based emissions-reduction targets. The horizontal axis represents time, going out into the future. Along the vertical axis, we see the percentage change in greenhouse gas emissions. As you can see, despite declining greenhouse gas intensity (reductions in the amount of greenhouses gases produced per unit of GDP), GDP is still rising at such a rate as to cause overall absolute emissions to increase. As a result, intensity-based targets might not lead to the prevention of the full range of climate change impacts. On the positive side, intensity-based targets could be an important first step to demonstrate that economic resilience

Figure 6.2. Hypothetical Relationship between GDP, Total Greenhouse Gas Emissions, and Emissions Intensity over Time

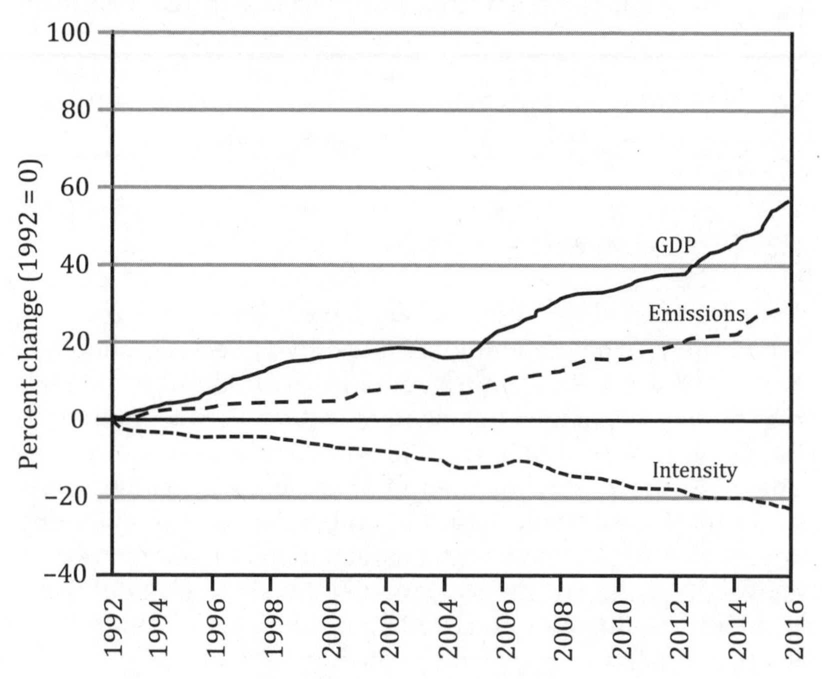

does not have to be sacrificed as we address climate change, and that it might also trigger the beginnings of a transition to renewable energy.

A further distinction to be made between various greenhouse-gas-reduction targets is that of targets that address a particular greenhouse gas such as carbon dioxide versus those that capture "equivalent carbon dioxide" (CO_2e). Equivalent carbon dioxide is simply a way of translating the global warming potential of various greenhouse gases into the equivalent effect these gases would have on the atmosphere if they were carbon dioxide. Think about it this way. Calories in food affect our weight, so they are a useful measure of the effect of various foods. An apple is around 50 calories, while an avocado is around 150 calories. If we thought in terms of "equivalent apples," then eating one apple and one avocado is like eating four apples worth of calories. So, if we say that our target is to reduce greenhouse gases to 350 ppm/volume CO_2e, this means we want to reduce the total stock of greenhouse gases in the atmosphere to the *equivalent* of 350 ppm/v CO_2, but we could this by reducing methane, nitrous oxide, or chlorofluorocarbons.

6.2. The global energy system

Over the past three decades, we have seen the far-reaching implications of a petroleum-based energy system. Oil shocks in both 1973 and 1979 were triggered by geopolitical maneuverings on the part of the Organization of Arab Petroleum Exporting Countries (OAPEC), the United States, and others whose interests were deeply invested in maintaining a steady supply of oil from the Middle East. Oil embargoes and the faltering supply that resulted from military action led to dramatic shortages around the world, accompanied by sky-rocketing prices that had a lasting effect on economic security for years following the events that triggered the shortages.

More recently, escalating oil prices have made possible the production of ever more costly and labor-intensive types of petroleum, such as the extremely dense bitumen-rich sands of Alberta and Venezuela. Previously thought to require too vast an input of resources for too small a yield, oil sands are now considered "unconventional" but nevertheless feasible sources of petroleum. The result has been great controversy over the environmental impact of oil sands mining, including the quantity of greenhouse gas emissions that are produced during the energy-intensive process of stripping the bitumen from the sand to which it is tightly bound, and the long-term implications of consuming this fuel at all versus accelerating a transition towards renewable energy.

Coal and oil provide most of the world's total primary energy supply, and the demand for fuel has nearly doubled in the past thirty years. In 1973, for instance, total world energy consumption was 6,107 million tonnes of oil equivalent (Mtoe). Of this total, 86.7 percent of energy was supplied by natural gas, coal, and oil. By 2010, we were consuming 12,717 Mtoe, 81.4 percent of which consisted of these three fossil fuels (Figure 6.3). The slight decline in the dominance of fossil fuels was due to the increasing importance of nuclear power, hydroelectric power, and an extremely small proportion of renewable energy (including solar, geothermal, and wind energy).[5] The direct result of our patterns of energy production is a spectacular increase in greenhouse gas emissions from fossil fuel combustion, which is only expected to increase in the absence of policies targeting the core issue: our energy supply.

These issues raise the question of the sustainability of our current global energy system. If our economy is fueled by a non-renewable resource that is becoming less affordable, less secure, and less environmentally sustainable with each passing year, where should we turn for energy to transport us from one place to another, drive our industry, and power our homes?

The sections that follow explore strategies at our disposal for reducing greenhouse gas emissions and mitigating climate change. These involve conserving energy through behavior change and energy efficiency technologies, as well as transforming our supply towards renewable, low-carbon sources of energy.

Figure 6.3. Global Primary Energy Supply by Fuel Type, 1973 and 2010

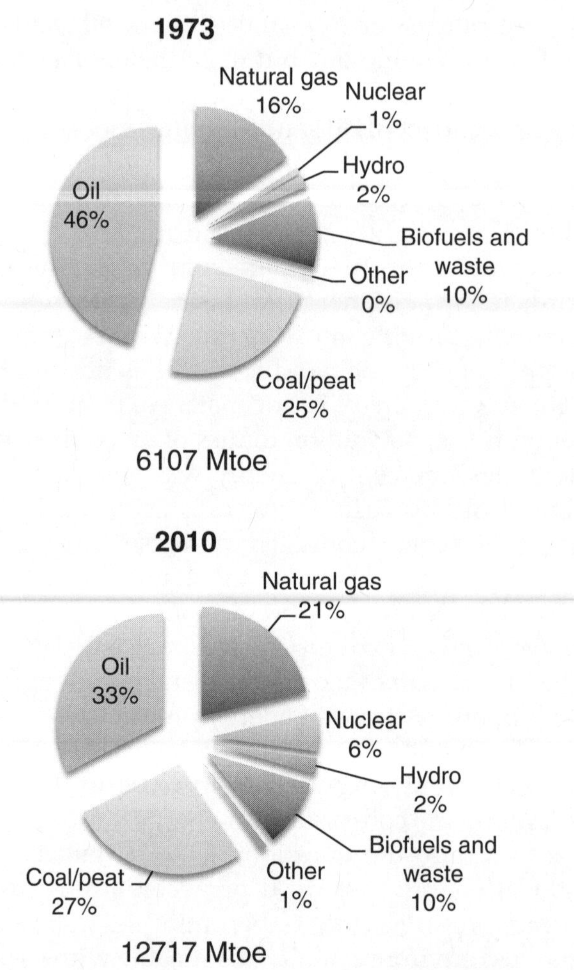

Source: International Energy Agency, *Key World Energy Statistics, 2011* (Paris: IEA, 2011).

6.3. Mitigation strategies

Since the combustion of fossil fuels is the main source of anthropogenic (human-caused) greenhouse gas emissions, it stands to reason that reducing our consumption of fossil fuels will lessen our impact on the climate. There are two main methods we can use to accomplish this goal: demand-side strategies, and supply-side strategies.

Demand-side strategies involve changing our behavior or our technologies so that we use less fossil fuel – for example, by better insulating our homes, turning off lights, or purchasing more efficient appliances. Supply-side strategies, in contrast, involve the use of completely new sources of energy to reduce our reliance on fossil fuels. This might include, for instance, producing electricity from wind power rather than from coal-fired power plants.

6.3.1. Demand-side mitigation: Energy efficiency and conservation

Reducing demand for fossil fuels that produce massive quantities of greenhouse gases can be attained through either energy efficiency or conservation. Energy efficiency refers to reducing the amount of energy needed to perform a particular task, such as lighting a room or transporting ourselves from home to work.

A simple formula that helps us understand this is to divide the service provided – such as maintaining a 2,500-square-foot (232-square-meter) home at 20°C for one year – by the amount of energy required to provide the service – say, 110 gigajoules of natural gas. If we compare this calculation to a home with a more efficient furnace, better insulation or other elements of home efficiency that requires only 80 gigajoules of natural gas to heat to the same level over the same period, the latter house is clearly more energy efficient than the former.

Energy efficiency usually increases with wealth, given that financial resources and high levels of education are often required to develop and roll out energy-efficient technologies. Along with wealth, however, have come higher levels of greenhouse gas emissions. That is, North Americans use less energy per unit of GDP than people in South Asia or sub-Saharan Africa, but North American GDP is so high that the end result is nevertheless comparatively high levels of greenhouse gas emissions. So, despite the higher per capita greenhouse gas emissions often associated with high levels of wealth, higher levels of *energy efficiency* are also found at the highest end of the wealth spectrum.

6.3.1.1. Energy-efficient technologies

Privately owned and operated vehicles are a core component of Western culture. Cars are associated with freedom, mobility, convenience, and luxury, and many cities, especially in North America, are built specifically with transportation via car in mind. These deeper, collective values and often-invisible habits[6] reinforce our dependency on the automobile, and create inertia with respect to transportation systems. The inevitable peak and then potentially rapid decline of conventional oil supplies – as well as the impacts of climate change resulting from the use of fossil fuels to power vehicles – has intensified the need for modes of personal transportation (and mass transit) that make better use of more energy-efficient substitutes. Indeed, the adoption of such alternatives could be one component of a multi-pronged approach to climate change mitigation that helps reduce air pollution, manages greenhouse gas emissions, and makes cities more livable.

Several strategies can make vehicles more fuel efficient. Different materials can be used in their construction, such as lighter composite materials, rather than heavier metals. Improved aerodynamics can help to reduce drag and minimize the work done by the engine. New technologies, such as plug-in hybrids, can pair cleaner electricity – if produced through hydroelectricity or some other renewable source, rather than from coal – with gasoline-consuming internal combustion engines, thereby significantly reducing the use of fossil fuels. Advanced tires and general vehicle maintenance also contribute to significantly higher vehicle efficiency.

Although the ability of vehicle engines to transform fuel into work has increased over the past twenty or thirty years, this increase in efficiency has not necessarily translated into decreased fuel consumption by standard North American vehicles. In fact, the gains made by increasing efficiency have been eaten up by the building of larger, more powerful vehicles (such as sport utility vehicles), which consume as much fuel as, or even more than, smaller, less fuel efficient cars.

Our homes and offices also present important opportunities for enhanced energy efficiency, and thus climate change mitigation. Older homes are often plagued by poor design that allows the escape of heat and poor use of electricity for light, but newer green building standards show dramatic progress towards energy efficiency. The core elements of building construction that might contribute to climate change mitigation are:

- the building's location and surroundings;
- the types of materials used in its construction;
- the placement and type of windows and doors;
- the color and design of the roof; and
- the devices (such as furnaces, lights, and other appliances) that are used within.

Ambitious and effective climate change mitigation strategies will consider ways to ensure that new buildings follow the state of the art in green building construction – through, for example, alterations to building code legislation. But regulations are also required that accelerate the uptake of energy-efficient building retrofits, since new construction represents only a small slice of the current building stock, and the operation of existing buildings is a major source of greenhouse gas emissions.

6.3.1.2. Conservation and behavior change

Now that we have learned a bit about energy efficiency as a demand-side mitigation strategy, let us consider the other side of that coin: conservation. Whereas energy efficiency requires that new technologies such as high-efficiency furnaces or hybrid vehicles be developed, conservation requires that we change our energy-consuming behavior. This brings to the fore a new constellation of challenges associated with the complexities of human behavior and the variety of strategies required to change it.

Early models of pro-environmental behavior tended to argue that knowledge about the environment fed directly into attitudes towards the environment, which, in turn, contributed to environmentally desirable behavior. Often termed "deficit models," these theories have been roundly criticized for their implication that behavior can be changed simply through education,[7] and for the assumption that science provides objective truths, which the public fails to understand.[8] A more obvious theoretical handicap is the distinct lack of empirical support for clear causal links between attitude and behavior.[9]

As a result of these criticisms, a more sophisticated model of individual behavior has been developed that incorporates more nuanced interpretations of the relative influences of beliefs, values, knowledge, attitudes, and behavior. According to this model,[10] factual knowledge, values, and attitudes directly influence intentions, which are the immediate precursor to behavior, and social pressures affect the degree to which individuals intend to engage in environmentally sensitive behavior. These shifts in social and behavioral psychology have deeply influenced traditional economic assumptions that individuals are purely rational and seek only personal benefits from their actions, leading to the increasing importance of norms, emotional drivers, and social pressures in the field of behavioral economics. Consider, for instance, the social pressure that still exists in some circles to smoke cigarettes. Although there is overwhelming evidence that smoking leads to coronary heart disease and lung cancer, and although education about these risks has been widespread for decades, millions of people still choose to smoke. Why? The reasons vary, of course, but include the perception that smoking is relaxing, attractive, subversive, or a way to connect socially with others. This is the social psychological context of behavior – crucial elements to consider as we explore action on climate change.

Despite attitudes and beliefs that might lead an individual to intend to behave in a "pro-environmental" manner, however, situational factors such as economic constraints, socio-cultural constraints, and even climate, might have a significant effect on the extent to which the individual actually behaves in a manner that is consistent with his or her attitudes, values, and beliefs.[11] For example, I might believe that driving my car has negative environmental consequences, and feel responsible for taking individual action, but if efficient and convenient public transit is not available to me, I am stuck. Similarly, I might feel social pressure to keep a car to demonstrate wealth or status.

One of the key factors that shapes and motivates our behavior is education. When it comes to the human contribution to climate change, it is important to understand the implications of undesirable behavior, the benefits of changed behavior, and the incentives that are available to help us make the shift. In our daily lives, in other words, we must understand that, for example, driving inefficient vehicles is contributing to the problem, and there are health benefits to choosing other modes of travel, such as walking or cycling. Education alone, however, is insufficient to change behavior. We also must have access to infrastructure that supports new behavior – such as safe and attractive paths on which to cycle to work, and showers and lockers to use once we get there. As well, individuals need incentives to behave in a more environmentally sensitive manner. These incentives could take the form of taxes – for example, on fuel, to dissuade us from driving frequently – or, as became the case with smoking over time, changed social attitudes about driving to work, and the positive reaction of peers to our decision to cycle, use mass transit, or car-pool, rather than drive alone.

At the same time, the practicality of alternative modes of transportation for many people is heavily influenced by the way cities, especially in North America, are designed, which forces heavy reliance on cars. As a result, transportation is responsible for 28 percent of greenhouse gas emissions in Canada, for example,[12] and vehicle ownership is double what it was in 1960.[13] Behavioral choice is also affected by the availability of alternatives, individuals' willingness to explore the potential of living without a car, and their understanding the implications of transportation that draws on unsustainable fossil fuels. This is a much more complicated set of factors than merely replacing an inefficient car with one that performs exactly the same service using less fuel.

In sum, demand-side mitigation offers significant potential to reduce greenhouse gas emissions and to begin to address the core causes of climate change. But we also need to make the transition to low-carbon sources of energy – in other words, supply-side mitigation.

6.3.2. Supply-side mitigation

Rather than focusing on using fewer fossil fuels through greater efficiency or conservation, supply-side responses address the critical issue of the sources of energy from which we draw. At the core of supply-side climate change mitigation is the use of renewable energy, which could fundamentally transform our energy system and alter our greenhouse gas emissions trajectory.

Fossil fuels are created over millions of years as organic materials decompose and are compressed under sediments. New supplies of conventional and unconventional oil are continually being discovered, though at a diminishing rate, and the supply of fossil fuels does not replenish itself naturally (recall Figure 2.5) at a rate that is economically useful. Renewable energy, in contrast – which recent estimates suggest fills approximately 19 percent of current energy demand – draws upon virtually inexhaustible sources of energy such as that from the Sun, "new" renewables such as small-scale hydro, tidal power, geothermal, wind, and biofuels, and more traditional renewables such as large-scale hydroelectric and biomass combustion[14] – both of which have significant environmental and social implications. For example, although forests are considered a renewable resource, biomass consumption can contribute to deforestation and unhealthy indoor air quality. So "renewable" is not always synonymous with "sustainable."

There are many benefits to implementing renewable energy technologies. They not only offer the potential to reduce carbon emissions dramatically and manage climate change; they also offer the promise of a more diversified energy source that is not tied so closely to unstable geopolitical realities. In developing countries, technologies such as solar cook stoves, which replace the combustion of dirty coal or biomass, could improve indoor air quality significantly, and contribute to better health and well-being. As well, local sustainability might be enhanced by the use of sources of energy that are locally plentiful rather than those that are transported at great cost over vast distances. Finally, while the traditional energy system is often based on highly centralized production (such as a single coal-fired power plant that services thousands of homes and businesses), renewable energy often has the potential to be modular and decentralized. District energy systems, for example, can be designed that are self-sustaining and removed from the grid, while individual homes could be powered through solar panels or micro-turbines. Such changes make renewable energy systems particularly resilient in the face of extreme weather, political instability, and variations in resources flows.

Even renewable energy, however, has its drawbacks. Some forms, such as large-scale hydro, can lead to extensive ecological damage – for instance, it is estimated that 60 percent of the lengths of the world's large river systems are moderately or highly fragmented by dams, interbasin transfers, and withdrawals for irrigation. As well, the cost of the transition from the current system, in which considerable resources have been invested, to renewable energy systems, can be quite high. Many renewable energy technologies are rather new and have not had time to take advantage of the economies of scale that come with widespread production and application. Related to this is the issue of the cutting-edge skills and technologies required for highly efficient use of, for example, solar, tidal, and wind power, which creates high hurdles for many individuals and jurisdictions wishing to make the transition.

Renewable energy faces two more challenges, which are also a critical distinction between renewables and non-renewable fossil-fuel-based energy. First, solar, wind, and tidal power are, as we shall see, intermittent sources of energy: the Sun sets, winds do not always blow, and the tides are periodic, meaning that we cannot always access the energy they produce when we want it. There are ways around this intermittency, including storing the energy gathered from renewable sources in a battery or other storage device, although the capacity of such devices deeply influences the ease with which renewables can support demand during peak hours or particularly cold months. The second challenge renewables face relates to the crucial metric of power density: the amount of power (work done over time, or a flow of energy) produced per unit of area (such as a piece of land with solar panels on it), or mass, or volume (such as a liter of oil). A reasonable natural gas well, for instance produces around 28 watts of energy per meter squared (W/m^2), while solar photovoltaics can expect to produce only around 6.7 W/m^2. Of course, real estate is not the only determining factor when we consider which resources to cultivate – clearly, environmental and social effects matter as well – but it might dominate investment decisions in the absence of regulatory, social, or political pressures in favor of renewables.

Without covering renewable energy sources in detail, it is helpful to look at a few examples and at the costs and benefits of these technologies as we pursue the transition from non-renewable to renewable energy.

6.3.2.1. Wind power

Wind is created by the differences in pressure that result from uneven heating of the planet's surface by the Sun. Air rushes away from areas of higher pressure (cold portions of Earth's surface) and towards the warmer, low-pressure portions. The generation of mechanical and kinetic energy from wind is an ancient practice that has been crushing grain and transporting goods for millennia. Only recently, however, have wind technologies evolved to capture efficiently the power of wind and transform it into electricity. The modern wind industry began in 1979 with the serial production of relatively small wind turbines by Danish manufacturers. Wind power is now the largest component of "modern" renewables generation capacity, which excludes traditional biomass and large-scale hydro projects. China is leading the world in terms of current installed wind power capacity and is second only to the US in terms of new capacity brought on line in 2012.[15]

Wind power is a highly attractive source of renewable energy because it is almost completely clean in its production (with the minor exception of the relatively simple materials required to build turbines and other wind-powered devices), and it is virtually inexhaustible. Wind power can be harnessed using very simple or very complex technologies, making it available to poor and rich alike and in both remote and central locations around the world. It can be used in a highly distributed fashion, such as with small turbines disconnected from a major power grid, which provide power to a single building or small community. Wind power can also be gathered from massive farms of turbines that feed electricity into a grid, powering entire cities or industrial complexes.

The core challenges associated with wind power are related to its intermittency and ecological impacts. Wind is a highly unpredictable phenomenon, and often does not coincide with peak electricity demands. Thus, the electricity produced by wind must be stored, which is an ongoing challenge for the industry. As well, complaints are often raised with regard to the danger wind turbines pose to birds, especially in ecologically sensitive zones such as flyways or breeding grounds. Controversy has also arisen surrounding the visual impact of wind farms on the landscape, as well as the "noise pollution" created by the turbines, leading some communities to reject the location of turbines near residential areas.

6.3.2.2. Solar power

Solar power is another "modern" renewable that is gaining traction around the world. Solar power is simply the generation of electricity or heat from sunlight. Photovoltaic panels can be used to collect and transform solar radiation into power, or the Sun's energy can be focused to boil water, which is then used to provide power. Like wind, solar power is intermittent, but predictably so, making simpler the task of anticipating when it will be available. Some solar experts argue that even the tiny percentage of the Sun's output that touches Earth is plenty to meet the energy and power needs of the entire human population many times over. The trick, however, is in capturing it efficiently and storing it effectively. Grid-connected solar photovoltaics are the fastest-growing energy technology in the world, with an average annual growth rate of 60 percent per year from 2007 to 2012. Germany added the most new solar photovoltaic capacity in 2012, and it also has the most existing solar capacity.[16]

The benefits of solar power are clear: it is an inexhaustible and completely clean source of energy – at least, once the solar panels are built, as energy and some toxic chemicals are used in their production. It is equally available in remote areas and poorer regions as it is in cities or highly developed countries. There is no need to process it beyond the amount that a solar cell does, unlike fossil fuels, which need to go through an extensive mining and refining process. Also important is the capacity for solar energy to be used in simple ways, such as with solar hot-water systems, in which tanks of water are simply located so as to be warmed by the Sun, and in complex ways, as with multi-megawatt solar thermal-concentrating power stations.

Challenges associated with solar power include concerns about the toxicity of chemicals and heavy metals used in the production of some photovoltaic cells, the availability of these metals in the future, the expense and technical sophistication of highly effective solar cells, and barriers to integration with other components of the current energy system.

6.3.2.3. Biomass and biofuels

The combustion of organic matter to yield heat is the oldest use of renewable energy sources, prevalent since the earliest human civilizations, and continues to provide heat and power for approximately 9 percent of global energy demand. Biomass is considered renewable because it causes no net increase in carbon dioxide and other greenhouse gases in the atmosphere (unless fertilizers are used). That is, the amount released through combustion is the same that the plants took out of the atmosphere as they grew.

Biofuels, in contrast, represent a more recent effort to transform vegetable fats, plant waste resulting from agricultural or industrial processes, and fermented sugar products into a highly efficient source of fuel. The growth of the biofuel sector has been driven by skyrocketing fossil fuel prices and concerns about climate change.

Biofuels are created and consumed through a cyclical process. Plants use solar energy and carbon dioxide to create carbon-based cellulose – the dense material that makes up all parts of a plant. These are harvested and processed to strip the cellulose from the plant matter, which is then broken down into sugars by enzymes. Microbes feed on these sugars, causing them to ferment and produce ethanol (a form of alcohol). Ethanol is a relatively clean-burning fuel, which can be used as a gasoline additive or can replace it entirely. A significant amount of heat is required to ferment sugars to produce biofuels, however, which runs the risk of creating an unsustainable cycle of production if clean sources of energy are not used to produce this heat. Biodiesel, another form of biofuel, is created by processing fats or oils with methanol (or ethanol) and sodium hydroxide. The result is a product that is much like traditional diesel and thus can be used in any car engine that normally consumes diesel.

Ethanol and biodiesel are the most widely available biofuels, but recent advances in the biofuels field have led to extremely high efficiency biofuel combustion methods, the creation of new forms of clean-burning mixtures such as syngas, and enhancements to biofuel production facilities to make the process more sustainable. An example of this last advancement is the use of biogas, a byproduct of landfills and organic matter decomposition, to produce the heat required by biofuel production processes.

The benefits of biomass and biofuels include the fact that no net carbon dioxide emissions are created (unless production facilities run on unsustainable sources of fuel), the byproducts of biofuel creation are relatively non-toxic and thus do not create challenges associated with spills and leakage, and biofuels integrate seamlessly with our current energy system.

These benefits, however, have not precluded significant debate over the value of biofuels. Traditional burning of biomass creates dangerous levels of particulate matter in the air, and thus has serious implications for human health. As well, cleaner use of biomass, as with biofuels, is not without challenges. In some countries, the value of biofuels is such that plant matter that could be used as food for humans is being diverted to the biofuel production process. This is the core element of the "food or fuel" debate. Similarly, ecologically sensitive lands are being cleared of trees and plant matter to feed biofuel production, leading to ecosystem impacts. In addition, issues remain concerning the quantity of water required to produce the feedstock and to supply the biofuel production process, and the inherent sustainability of biofuel production systems that must draw on fossil fuels for heat and power. Despite these criticisms, however, biofuels are often viewed as a useful "transitional" source of renewable energy that could help to fuel our current fleet of vehicles and industrial systems while other renewables become cheaper and more effective.

6.3.2.4. Geothermal energy

Geothermal energy is power extracted from the heat within Earth. This heat results from radioactive decay processes that continually take place in the planet's mantle – the middle layers between the upper crust and the core – and from the collisions of material that occurred during the original formation of the planet. Geothermal energy has been used to provide heat for thousands of years, but by drilling deep into the ground, often into hot aquifers, the much higher temperatures found there can be harnessed to produce electricity.

Evidence of the prevalence of accessible geothermal power is found in the hot springs that are located in every continent, including Antarctica, but it currently accounts for only about 0.3 percent of the world's electricity consumption. Only twenty-four countries take advantage of this source of energy; the largest group of geothermal power plants is in the United States, but El Salvador, Kenya, the Philippines, Costa Rica and Iceland use geothermal power for more than 15 percent of their domestic electricity needs.[17]

Geothermal energy is also used by simple ground-source heat pumps, which create a gradient in which comparatively warm ground (about 2.5 meters below the surface) causes heat to flow through a heat-exchange fluid towards the interior of a building, while the cool air inside the building is transported through the fluid back to the ground (and then warmed again). More than 2 million ground-source heat pumps are used in over 30 countries around the world.[18]

Geothermal energy is a viable source of renewable energy because it is widely available and requires no fuel to produce. The heat extracted through geothermal systems is tiny compared with the amount produced by the planet's core or by solar heating of the crust, and is thus considered environmentally sustainable. Geothermal power is also highly scalable – that is, it can be used to provide heat or electricity to an individual building or to an entire city. As well, it can feed directly into the existing electricity grid, making it relatively simple to integrate with current energy systems.

Relatively few challenges are associated with geothermal energy. Large-scale geothermal installations have high capital costs, which puts them out of reach of poorer communities or nations. In a very small number of cases – such as a demonstration project in Basel, Switzerland – earthquakes have been triggered after a geothermal well was drilled deep into a fault line. As well, although geothermal power is considered globally sustainable, long-term withdrawal in a single location might exhaust the local capacity, so extraction must be carefully monitored to ensure continual supply. In some countries, such as Canada – the only country along the Pacific Rim of Fire that does not exploit geothermal energy – arguments against geothermal include questions of the cost effectiveness of current technology, the abundance of hydroelectric power, and the inability to export the power, unlike natural gas or coal, for instance.

6.3.2.5. Tidal power

Tidal power is a form of hydroelectric power that derives directly from the gravitational forces among Earth, the Moon, and the Sun, combined with Earth's rotation, which drive the movement of Earth's oceans. Turbines can capture the energy embodied in this movement in much the same way as they capture the energy held by water that is pulled by gravity down a slope, as in the case of more traditional hydroelectric power systems.

Tidal power systems make use of the kinetic energy of moving water to power turbines that generate electricity, and have relatively little impact on ocean ecosystems. Barrage systems, in contrast, make use of the difference in the height of the high tide versus low tide, and function like dams that power turbines as the water is allowed to flow from the higher elevation to the lower. Tidal lagoons function essentially like barrage systems but at a smaller and less ecologically impactful scale.

The world's first deep-sea tidal energy farm was announced in 2007, and will be constructed off the coast of Wales. Advances in turbine technology are making tidal power a more feasible renewable energy option around the world, although much progress must be made before it provides a significant portion of global energy demand.

The benefits of tidal power are similar to those associated with other forms of renewable energy. Tidal power produces no greenhouse gases – except potentially in the initial production of the system materials – and thus can be an important component of supply-side mitigation strategies. Sites for small-scale tidal power production can be found around the world, often in rather remote areas. Although intermittent, tides are predictable in their schedules, and so facilities can plan for their ebb and flow.

Barrage systems, like large-scale hydro, is the tidal technology with the greatest ecological impact. The vast size of barrage systems creates dramatic changes in the characteristics of the water on either side of the barrage, altering the ecosystems that existed before the intervention. A smaller amount of water is exchanged between the barraged basin and the sea, reducing the turbidity (or amount of particulate matter in suspension) of the water and creating conditions in which phytoplankton can flourish. This alters the productivity of the ecosystem, and changes the conditions in which the species in this habitat live. Fish mortality can occur as a result of all three types of tidal power systems, and the variability and intermittency of tides create challenges for the storage of energy. Finally, high infrastructure costs plague large barrage systems, and few sites in the world have the proper geographic characteristics to support this kind of development.

The renewables explored above are not the only ones at play, of course. Large-scale hydroelectric power, for instance, is the most commonly used form of renewable energy, providing more than 16 percent of global electricity needs,[19] and a crucial source of low-carbon energy. There is growing opposition, however, to the massive inundation and habitat degradation that is an inevitable part of major dam construction, as well concerns about the resources consumed in making the dams. Nuclear energy is an equally contentious issue, dividing proponents of action on climate change. Some argue that the rapid uptake of nuclear power, given proven technologies, available materials, and cost effectiveness, is the only viable strategy for breaking free of the shackles of fossil fuels in the near term. Others argue, especially in light of the Fukushima Daiichi nuclear disaster in 2011, that the magnitude of the risks and interminable legacy of nuclear mishaps are reason enough to turn to other forms of renewable energy.

Taken together, solar, wind, geothermal, biofuels, and tidal power represent crucial elements of a transition to renewable energy and the supply-side mitigation of climate change. One important final note is that demand-side and supply-side mitigation are *complementary* strategies. Reducing our demand for energy, no matter what the source, can limit the strain on power-production facilities during peak consumption hours. This, in turn, might reduce the need for investment in power plants, and facilitate the integration of renewable, though more intermittent, sources of energy.

The final ingredient of traditional mitigation is the capture and storage of carbon and the enhancement of carbon sinks, to which we turn next.

6.3.3. Carbon capture and storage

Choosing cleaner fuels or using less of them are not the only mitigation strategies at our disposal. Carbon capture and storage and carbon sequestration also have the potential to contribute meaningfully to solving the climate change problem. Both carbon capture and storage and carbon sequestration are, in effect, ways to make the drain in the tub bigger, rather than to decrease the flow of water into the tub by lowering emissions.

6.3.3.1. Carbon capture and storage

Carbon capture and storage (CCS) consists of a three-part process: (1) the collection of the carbon dioxide that is produced through industrial processes or fossil fuel combustion; (2) the transport of this CO_2 from the source to a suitable storage location; and (3) the sealing of the storage capsule so that the CO_2 cannot reach the atmosphere. There are two major types of CCS. The first type, orchestrated by humans, uses non-biological mechanisms to prevent the interaction of CO_2 with the atmosphere; although it is being seriously tested around the world, it remains one of the less common mitigation strategies. The second type, carbon sequestration, involves the absorption and storage of carbon by living organisms.

Non-biological CCS is widely regarded as having significant potential as a mitigation strategy, but it cannot solve the climate problem single handedly. As one element of a portfolio of mitigative actions, CCS can serve as a "bridge" between our current fossil-fuel-based system and a lower-carbon future. Because of its applicability to current technologies and fossil-fuel-based energy sources, CCS can help reduce the amount of greenhouse gases we are putting into the atmosphere while we search for and refine zero-carbon technologies.

Although some electricity is produced through renewable energy sources, the burning of coal remains the dominant source of electricity in the world. As such, one of the most promising applications of non-biological CCS is in electricity-production facilities, where emissions can be reduced by as much as 90 percent, although the benefits are partially offset because a plant employing CCS consumes more energy – and thus would typically produce more greenhouse gas emissions – than a plant without CCS capabilities.

Since the costs of capturing, transporting, and storing CO_2 are so high, CCS technology must be applied to large industrial facilities that produce massive quantities of greenhouse gases. Our current energy system is based on the large-scale, centralized production of electricity and heat, so it conveniently provides many large single sources of CO_2 (or point sources) that could be part of a CCS system. For instance, CCS could be applied to the production of energy from biomass (such as wood waste), coal-fired power plants, and synthetic gas production facilities, to name but a few. In fact, this is what is meant by the phrase "clean coal": applying CCS to coal-fired power plants to reduce or eliminate the emissions produced at these facilities.

Transport of CO_2 links the source of the gas to its storage facility. Although technologies exist to transport CO_2 in solid, liquid, or gas form, there are trade-offs associated with each. Massive amounts of energy, for instance, are required to transform CO_2 from a gas into a solid. As a gas, however, CO_2 is already being transported across the United States and Canada via pipelines that were constructed to use CO_2 to enhance oil recovery by injecting it into oil wells and pushing out usable oil.

A variety of options exists when it comes to storing CO_2. The most commonly discussed option is to store CO_2 in geological formations, such as exploited oil and gas reservoirs, coal beds of no commercial value, and deep saline aquifers into which the CO_2 will dissolve. In no case is CO_2 injected into vast empty caverns; rather, it is stored in porous sediments or liquids.

Another storage option is to inject CO_2 deep into the oceans, where it would remain isolated from the atmosphere for millennia. Over time, this CO_2 would gradually exchange with the atmosphere, and the end result would be the same as if the CO_2 had been injected directly into the atmosphere. The difference, however, is the time scale over which this would occur. The near-term dangers associated with this storage strategy are mainly ecological: the ocean in the direct vicinity of the CO_2 injection site would be acidified, causing potentially dramatic ecosystem changes.

The final way to store CO_2 after it has been captured is through the transformation of the gas into stable solids. In this case, CO_2 is reacted with metal-oxide-bearing materials, yielding carbonates that would isolate CO_2 from the atmosphere for millennia. Unfortunately, this is an energy-intensive process – a full CCS power plant with mineral carbonation would require between 60 percent and 180 percent more energy than a typical plant without CCS – and in any case is not yet commercially viable.

Of all the CO_2 storage options, geological storage appears to be the most promising. We already have the technology to capture CO_2 from large industrial facilities and, as noted, in some cases pipelines are already in place to transport and store CO_2 in depleted oil and gas reservoirs. This would become a more cost-effective option if a modest price were placed on carbon – for instance by adding a tax to fossil fuels, and requiring individuals and businesses to pay a little more to drive their cars, heat their homes, and produce goods and services. Ocean storage is less attractive, as there are many potential risks to ocean flora and fauna. Mineral carbonation is the least-developed CCS technology, as it requires massive inputs of energy.

The controversies surrounding non-biological CCS are many. Some fear a sudden release of highly concentrated CO_2 from underground reservoirs, although there is little evidence that this is likely. More certain are the ecological effects associated with ocean storage. Finally, some argue that employing CCS only allows us to continue using fossil fuels, which inevitably will run out. Thus, CCS is at best a bridge from a high-carbon system to a low-carbon one. At worst, it prevents us from developing truly low-carbon technologies and creating real solutions to the climate change problem.

6.3.3.2. Carbon sequestration

Like carbon capture and storage, carbon sequestration involves taking carbon out of the atmosphere after it has been produced through the combustion of fossil fuels or the decomposition of organic matter. In contrast to CCS, however, carbon sequestration involves the capacity of photosynthesizing plants – trees, crops, grasses, and even algae in ponds and oceans – to absorb carbon from the air, binding it into their plant tissues. This mitigation strategy is of great interest to countries with vast quantities of forested land, such as Canada, the United States, and Brazil.

Two core strategies can be used to enhance the amount of carbon that a piece of land sequesters. One, afforestation, involves planting trees on land that has never (or at least, not in recent memory) had trees on it, causing a net increase in the amount of carbon that is bound into that piece of land. The other strategy, reforestation, involves replacing forests that were recently removed.[20]

A key problem with carbon sequestration as a mitigation strategy is that forests are not permanent. A forest fire could (and often does) tear through the area, combust the living matter, and release the CO_2 bound in the plant tissues back into the atmosphere. As well, humans could use the wood for fuel, or clear the land for agriculture or other activities. Issues of land management also arise. A particular portion of forest might be preserved to sequester carbon, but an adjacent portion of land might be deforested instead, which would be especially damaging if the adjacent area is particularly ecologically sensitive or culturally significant. Finally, there is considerable uncertainty associated with carbon sequestration. Tree species absorb particular amounts of carbon given a particular climatic state, but the climate is changing, which complicates the issue of calculating how much carbon will be taken up by a particular area of trees in the future.

As a result of these controversies, sequestration of carbon in forests, while ecologically and socially critical, is not viewed as an effective long-term mitigation strategy.

6.4. Fostering accelerated and transformative mitigation

In this chapter, we have explored the elements of the global energy system that are unsustainable, in part because they contribute directly to anthropogenic climate change. Mitigation, or addressing the causes of climate change, is a crucial dimension of sustainability, and can be pursued through either demand-side or supply-side strategies. Coordinated and accelerated implementation of renewable energy technologies has the potential to transform our global energy system, and even reduce the need for conservation or efficiency. Even so, this transition will take place in the context of complex institutions, deeply held values, and powerfully habitual behaviors. As such, the mitigation question is not simply an issue of technological development, but of creating the institutional rules and enabling environments that foster a change in values. This is at the core of transformative change, and requires technical as well as social innovation, issues that we discuss in greater detail in Chapter 11, as we look forward to a sustainable future.

CHAPTER SEVEN

Climate Models

MAIN POINTS:

- Climate models are grounded in physical principles such as conservation of mass, energy, and momentum, and laws of motion.
- Climate models are constrained by, and tested against, observations of the real world.
- Climate models help us understand how the different parts of the climate system behave and interact with one another.
- Climate models tell us about the probabilities of future climate scenarios, including both rates and magnitudes of change.
- The largest sources of uncertainty in models today concern clouds, aerosols, and feedbacks involving ice and ocean heat uptake.

How likely is Miami to be submerged by rising sea level within a human lifetime? What might changes in climate mean for rice production, freshwater supplies, or coral reef ecosystems? One way to answer these questions is to make thorough observations and record what happens over the coming decades, but with observations alone the answers might come too late. We have the desire, and the means to peer into the future, so that we might take action to influence future trajectories or ease transitions. Should we plant olives here? Should we abandon Amsterdam? How fast should we change our energy system?

In reality, we cannot simply try out a variety of possible futures to find out what happens to Miami or coral reefs, so we turn to models to help us learn. Models are attempts to represent the world. They can be physical (such as model airplanes), mathematical (such as computer models), oral (such as language), written (such as history textbooks), drawn (such as maps), and other forms. They can also be predictive. Perhaps the most important are mental models: those ideas in our brains that explain, to us, how the world works.[1]

We use mental models to forecast the future all the time. We decide whether we have time to start a load of laundry before the teakettle boils on the stove. If you have a dog, you probably keep a running model of his digestive system in your head that improves with experience. You can look up tide tables for your favorite beach far in advance. Modeling the future is not particularly unusual or magical.

Climate models are representations of Earth's climate system. They are usually mathematical models with equations that relate stocks and flows. The more complex climate models are dynamic, with feedbacks, lags, and evolution of the system over time. Using models, we can run experiments to isolate particular parts of the climate system, although, in the real world, we cannot turn anything on or off. For example, we can run a model that excludes the effects of water vapor, then add the water vapor variable and see what happens, or we can add or remove an ice sheet and see what happens. The results inform us about the magnitude of the water vapor feedback or the ice-albedo feedback. Using models, we can learn how Earth's climate system is likely to respond to different future scenarios and we can estimate the probabilities of certain outcomes, which allows us to narrow the range of those that are likely and eliminate others. The future is not completely unpredictable – some future pathways really are more likely than others. This information is key for community planning and for the implementation of mitigation and adaptation strategies.

In this chapter, we outline the different classes of climate models and their uses. We discuss what aspects of climate models are currently well known, what parts are more poorly known, and how people test models to increase confidence that the output reasonably represents Earth's climate. In Chapter 8, we will explore what these models, and other scenarios, tell us about emissions trajectories and climatic futures.

7.1. Climate model basics

Modeling systems requires us to do our best to represent the behavior of the relevant stocks, flows, feedbacks, and lags. We make judgments about what is relevant, based on the question we want to answer or the purpose of the model. A simple climate model, such as the energy balance model we describe later in this chapter, can be built in a couple of hours, but many scientists spend their careers building, testing, and improving sophisticated climate models.

7.1.1. Physical principles

At a minimum, building a climate model takes knowledge and skill in physics and math, with some accounting. Physics is key because climate models are grounded in physical principles. Models must comply with the requirements of conservation of mass and energy, laws of thermodynamics, and laws of motion in their representations of, for example, wind patterns and ocean currents. A model has to keep track of all the material and energy in the system it is modeling. Mass and energy cannot just appear or disappear in a model, or, if they do, modelers need a good reason for it. When a forest burns, the carbon contained in those trees goes somewhere – into the atmosphere as CO_2 or soot particles or into pieces of charcoal. A new addition of mass or energy to the system has to have a logical source. You cannot, for example, just double the Sun's energy output and then claim that your model is a realistic representation of today's Earth. If your model "uses" some energy to evaporate water (part of latent heat transfer), that same energy cannot also be counted as radiating away from Earth's surface (look back at Figure 2.6). A glacier model needs to keep track of the stock of water in the glacier as it grows or melts. This is where accounting skills matter. Your model will blow up if you cook the books.

As in any system, the parts have to be appropriately connected. If, in your model, atmospheric CO_2 goes up – because of deforestation or fossil fuel burning, for example – then the magnitude of the greenhouse effect must go up appropriately. To figure out what is appropriate, you would need to include the probability that the additional CO_2 molecules will intercept radiation from Earth's surface and the subsequent change in the stock of energy in the atmosphere. Then, that extra energy makes water molecules move around faster and increases the partial pressure of water vapor in the atmosphere, which, as a greenhouse gas, causes further warming. When the relative humidity in your model increases enough, some of the modeled water vapor should condense to liquid, decreasing the relative humidity and perhaps causing precipitation. In other words, your model must include relevant, physically grounded feedbacks.

7.1.2. The role of observations

Useful climate models are informed by real-world observations, which tell modelers what is realistic and what is out of bounds. By comparing the output of a model to observations, modelers can evaluate how well the model represents reality. Mismatches indicate areas where the model needs work. In turn, model results might prompt efforts to go out into the real world and make better observations.

Perhaps the best-known example of a combination of observations and models involves Earth's surface temperature. Researchers have used observations (from thermometers and satellites) to reconstruct global average surface temperature back more than a hundred years. Constructing this record is itself a non-trivial task, but different research groups, using different approaches and assumptions,[2] have produced essentially the same answer for historical temperature changes. A climate modeler's goal might be to represent realistically all known, relevant processes that influence the stock and flow of energy in Earth's climate system, producing, among other things, an estimate of global average surface temperature. The output then can be compared with real-world observations to see whether the model reproduces the measured temperature record.

If the model does not reproduce the observations, we must ask what is missing or what is not represented realistically enough. If the model seems realistic, however, one might choose to run it into the future with higher confidence that its output is reasonable. One might also use the model to ask more specific questions: how large an influence do clouds, water vapor, or aerosols, have on temperature, and what is the direction of their influence? what is the equilibrium climate sensitivity (see section 2.2.2.2.)? can we explain the observed historical temperature record without one or more factors? In 2007, the Intergovernmental Panel on Climate Change (IPCC) summarized an important result comparing observations and model output: natural processes alone – such as solar variability and volcanoes – cannot account for the twentieth-century temperature record, but the record can be explained if the activities of humans – such as fossil fuel burning and land-use change – are included.[3]

Beyond surface temperature, weather balloons and satellites measure conditions such as temperature and water vapor vertically through the atmosphere, which are crucial for checking models that simulate atmospheric circulation. A growing array of instruments (thousands of them) drifting in the oceans monitor temperature, salinity, and currents in the upper 2,000 m of the water.[4] These observations not only help us to estimate how much energy the oceans absorb of the current planetary energy imbalance (the answer is, a lot[5]); they also feed into (and test) models that simulate energy and mass flow in the oceans.

The interplay between observations and models has been particularly useful in tracking carbon through the Earth system. Only about half the carbon humans emit stays in the atmosphere in a given year; the other half goes somewhere else. To figure out where, scientists use atmospheric CO_2 measurements around the world (observations), combined with models of air flow (also informed by observations), which suggest how much of the carbon goes into the oceans and how much goes into land biomass at different latitudes.[6] These modeling results then have prompted further observations in land and ocean ecosystems, to quantify CO_2 uptake in different places. From such observation-and-model efforts, we now have better estimates regarding carbon fluxes, including how much of our emissions goes where.

Information about human activities that feeds into climate models is also based on observations. Fossil-fuel-extraction records and measurements of carbon isotopes in atmospheric CO_2 tell us how much fossil fuel burning contributes to atmospheric CO_2 concentrations. Observations of land-clearing obtained from satellites tell us about the rates at which we are altering Earth's surface. Over time, these observations help inform our thinking about the future. Given today's trends, what future pathways for human activities and climate are probable? What effects will our actions now have on those futures? Or, working backward, what future do we want to achieve, and how can we get there? There is more about this distinction between forecasting and backcasting in Chapter 8.

In short, models depend on observations. No near-realistic climate model stands alone, separate from any observed reality. Initial observations can be used to build initial models, which can inform future observations, which in turn can help refine the models. Over time, this back-and-forth helps improve our understanding of the climate system.

7.1.3. Time and space

Based on questions of interest, modelers make decisions about the spatial scales and time steps the model will include. For spatial scale, the modeler might choose to characterize Earth's oceans and atmosphere as two big boxes, or divide them up into many boxes or "grid cells" and model the transfer of mass, energy, and momentum among all the cells (Figure 7.1). In the model, each grid cell has average values for information of interest: temperature, relative humidity, cloud cover, wind speed, and so on. These change as the model runs forward in time, and mass and energy are exchanged among the grid cells. The model will recalculate all the values for all the grid cells for each interval of time the modeler chooses.

A model's time step is the interval between snapshots that show what the system looks like. Should one "look" at the system every second? every year? every thousand years? Modelers choose time steps that capture the processes of interest for the chosen spatial scale, which means that time and space choices are fundamentally related. If you are modeling, say, a teakettle, the processes of interest happen on a small spatial scale and over a short time. Your teakettle model should probably examine the state of the system every minute or so – it would make no sense to model the behavior of a teakettle using a time step of a year, unless it is a discarded teakettle rusting away. Similarly, if you are modeling the melting of the continent-sized ice sheets between the last ice age and today, you do not need to have a look every minute; you can choose a much longer time step, one more appropriate to the rates at which the processes of interest change. Space scales and time scales in a model need to align with one another.

Modeling choices involve trade-offs among the number of grid cells, the time step chosen, details of modeled processes, the total period modeled, and the number of model runs you wish to make. The more grid cells and detail in the stocks, flows, and feedbacks the model contains, the more calculations are needed to move the model one time step into the future. As with any computer process, each calculation takes real time, and a model with many boxes (Figure 7.1, panel B) and many items to keep track of in each box will have to make many more calculations per time step than a model with few boxes (Figure 7.1, panel A).

Figure 7.1. Modeling Choices: Levels of Detail

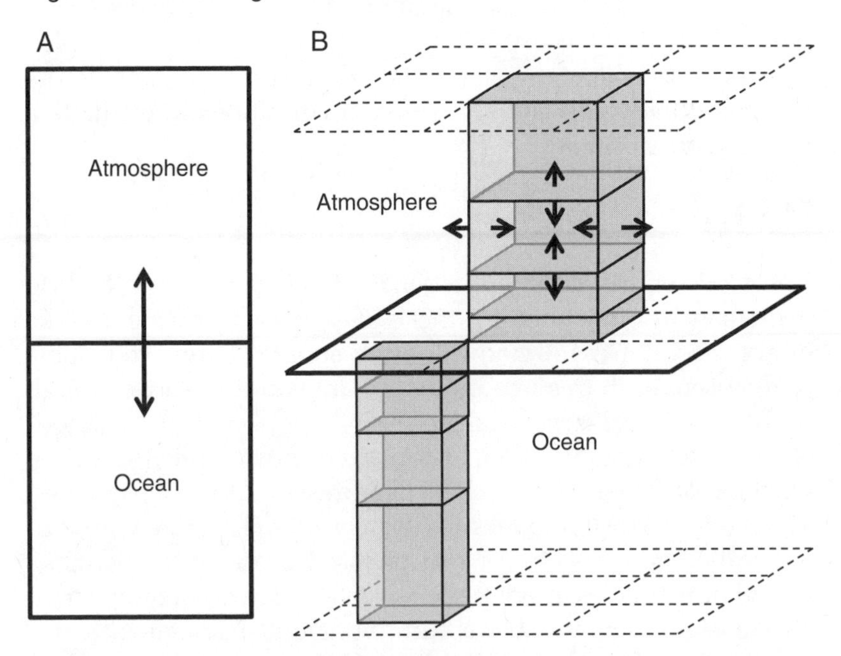

Panel A: A model might represent the atmosphere and oceans each as one simple box. In panel B, the model divides the atmosphere and oceans into many boxes, or grid cells, both horizontally and vertically. Arrows represent exchanges of matter and energy between adjacent cells. As well, imagine additional arrows towards and away from you, out of and into the page. Typically, the grid cells are thinner near the surface, and get thicker both higher in the atmosphere and deeper in the oceans. Model grid cell sizes in the oceans are typically smaller than in the atmosphere. The more grid cells, and the more variables tracked in each cell, the more computer power required.

More detail per time step means that the total number of steps you can afford to take might be limited by computer power: if you are taking a snapshot every minute, you probably could not afford to run your model out 10,000 years, nor would you want to. Generally, the smaller the time step, the less total time can be modeled.

Since their invention, the speed of computers has increased dramatically, and as they get faster, modelers have more options: (1) they can increase the resolution of their model, and make both the spatial boxes and time steps smaller (since these scale together); (2) they can run the model for a longer overall time period; (3) they can explicitly include more processes, which require more calculations; and (4) they can run their model multiple times with slightly different conditions to evaluate a range of possibilities.

7.1.4. Parameterization

As a modeler, are you interested in the forest or the trees? Do you want to include the photosynthesis of each leaf, the water uptake of every root? Do you know enough about the trees to be able to represent each tree in a model, or are you limited to a rough understanding of larger-scale forests? An aspect of climate models related to computation cost involves how explicitly, and at what level of detail, the model will represent different processes. Modelers choose to model some aspects of the climate with representations as close to the real physical processes as possible. For other processes, modelers choose to approximate important effects – using reasonable assumptions and relevant factors – rather than explicitly model in detail. These choices depend on the questions asked, computational limits, and sometimes the limits of our understanding of climate processes. Often, the processes approximated (rather than explicitly modeled) are those that operate in a smaller space or shorter time scale than the model, but have important aggregate effects that influence the climate at larger scales, like the model scale. Although each individual tree exchanges carbon with the atmosphere and soil, we might limit our model to estimating the full effect of the forest.

Approximating aggregate effects and using those approximations in a model is called "parameterization." A common example in climate models is to parameterize clouds. Clouds play a crucial role in Earth's climate, so leaving them out is not a realistic option, but typical cloud size is usually smaller than the grid cell size used in climate models. Plus, cloud formation depends on many microscale processes, such as condensation of water vapor. Explicitly modeling the formation of each raindrop, then modeling the raindrops' formation into clouds, is a level of detail that cannot realistically be included at present, so modelers use parameterizations. They approximate cloud cover in different grid cells using information relevant to cloud formation, such as air temperature and relative humidity, and they estimate the distribution of droplet sizes in clouds (which influences cloud reflectivity and precipitation), but they do not attempt to model individual raindrops.

Certainly, if one were interested in the processes by which raindrops form, and how varying different factors would affect raindrop formation, one could and should model those processes, using small-scale observations to constrain the model. Information from that small-scale work then can be incorporated into better parameterizations in large-scale models. But that level of detail is of questionable value in modeling the global climate. In addition, the model would never finish running – it would take too long. Choices regarding what to parameterize, and how, depend on the question of interest and the best efforts of the modeler.

7.1.5. Testing climate models

How can we evaluate climate models? Grounding models in physical principles and observations is a good start. Our confidence in a model increases if it reasonably produces something that looks like the real world. We can also investigate how well a model simulates past climates, for which we have evidence to check model output. These past climates might be data from the most recent century, or much older periods reconstructed from paleoclimatic evidence. If the models fail to reproduce observations, their forecasts for the future are likely suspect. But if they can reproduce the behavior of Earth's climate system in the past, we should have more confidence in their forecasts of the future.

No single climate model or single run of a climate model has all the answers. Different models use different assumptions in their parameterizations, modelers make different choices about what to model explicitly and what to parameterize, and physical features of Earth's climate system have ranges of variability. Modelers therefore use collections of model output – "ensembles" – to increase their confidence that the range of estimates from models is reasonable. One ensemble approach is to feed the same climate forcing – say, prescribed future human emissions of CO_2 over time – into different models, and then compare the range of outcomes.[7] Another approach is to run the same model many times, but with slightly different input values that span the range of likely variability based on physics.[8] Combinations of these – multiple models with multiple scenarios – are also used systematically to help constrain the likely range of outcomes.

7.2. Types of climate models

Mathematical climate models come in different degrees of complexity, depending on the questions asked. In this section, we examine one of the simpler types of climate model – energy balance models – in some detail. These models allow us to answer questions about overall stocks and flows of energy in, out, and within Earth's climate system. For more complex questions and purposes, we need more complex models, and so we also describe the basic categories of more complex models and the types of questions such models can answer.[9]

7.2.1. Energy balance models

Energy balance models (EBMs) are the simplest type of climate models, but they are still useful to gain an understanding of how overall energy flows behave and how changes in greenhouse gases, reflectivity, or energy from the Sun influence Earth's average equilibrium temperature.

We can build a basic EBM using just the Stefan-Boltzmann equation, $F = \sigma T^4$ (equation 2 in section 3.1.1.), and some additions and subtractions. First, assume Earth has no greenhouse gases and that Earth's surface perfectly absorbs all the incoming solar radiation (341 W/m^2, because there would be no reflection; see Figure 7.2, panel A). In this model, Earth's surface would have to radiate 341 W/m^2 to be in energy balance. According to the equation, Earth's temperature would end up at about 278 K (5°C). But what does the phrase "would end up" mean?

Even though an EBM has very few equations to deal with, and we can easily jump right to the "answer" – Earth's temperature, once the system is in balance – we can still think about what happens over time for Earth to reach that equilibrium temperature. Imagine turning off the Sun for a bit. The solar system would get dark and cold, and Earth would radiate some small amount of energy in accordance with its cold temperature. Now, turn the Sun back on. Earth starts to absorb solar energy and warms up, and as it does, it radiates more energy (see section 3.1.1.). For a while, inflow from the Sun exceeds outflow via Earth's radiation, and it would take some time for Earth to warm up to the equilibrium temperature at which its outflow matches the inflow. As long as inflow exceeds outflow, Earth will warm further. If, for some reason, outflow exceeds inflow for a time, Earth will cool down, and therefore radiate less, until inflow and outflow are back in balance. This is one of the most important stabilizing feedbacks in the climate system, which we can model with a simple EBM.

In Figure 7.2, look at the trajectory for Earth's temperature over time. At the beginning, there is a period during which inflow is greater than outflow, and Earth is warming up. Then the model output stabilizes at an equilibrium temperature of 278 K (5°C). Now, perturb the model by adding 30 percent reflection (equivalent to Earth's albedo today) and keep running the model forward in time (Figure 7.2, panel B). After some delay during which outflow exceeds inflow, Earth would stabilize at an average surface temperature of only 255 K (–18°C).

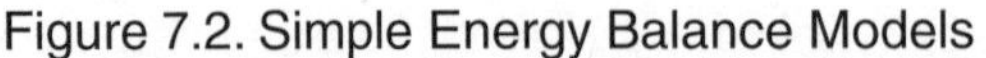

Figure 7.2. Simple Energy Balance Models

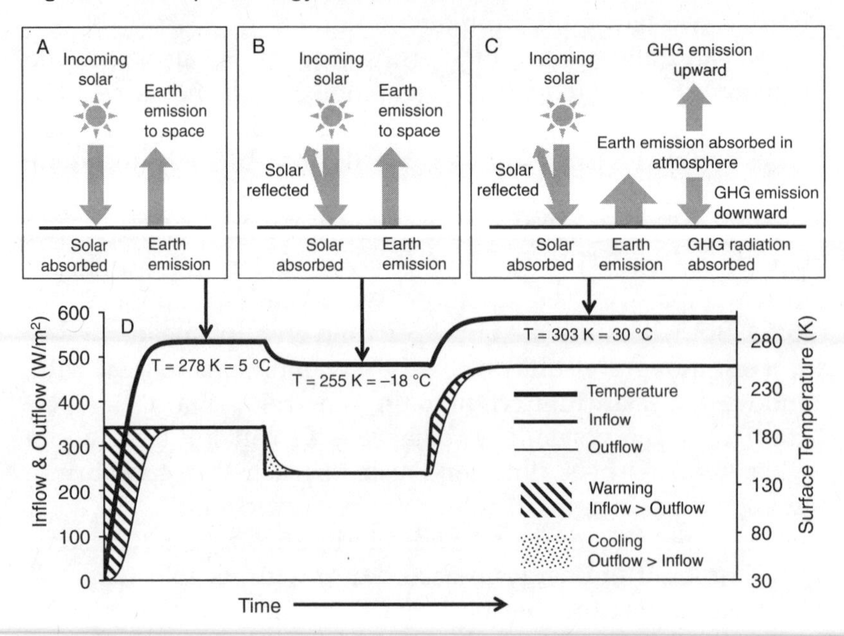

These models use the Stefan-Boltzmann equation, $F = \sigma T^4$, where F is the energy radiated in W/m^2, σ is the Stefan-Boltzmann constant, and T is the object's temperature in kelvin. In the model shown in panel A, there is no reflection and no greenhouse gases. Panel B shows the model in panel A with reflection added. Panel C shows the model in panel B with greenhouse gases added. Panel D shows the changes in temperature, energy inflow, and energy outflow over time as Earth warms up to equilibrium for the model in panel A, then cools off as the model in panel A transitions to equilibrium for the model in panel B, then warms again as the model in panel B transitions to equilibrium for the model in panel C.

Next, add greenhouse gases, and assume that (1) Earth's surface absorbs all the incoming solar radiation that is not reflected – in other words, the atmosphere does not absorb any of that radiation directly – and (2) the greenhouse gases absorb all the energy emitted from Earth's surface and then re-emit half of the energy upward and half back downward towards Earth (Figure 7.2, panel C). We know that, once equilibrium is reached again, the energy outflow from the upper atmosphere must equal the solar energy absorbed by Earth. We have set the energy flowing downward from the atmosphere to Earth's surface to equal the energy flowing upward, since greenhouse gases re-emit energy in random directions that average out to about half up and half down. Now, Earth's surface has two inflows, one from the Sun and the other from the atmosphere. At equilibrium, Earth's surface has to re-radiate all of the energy from *both* sources, so the temperature would have to be 303 K, or 30°C. Before reaching equilibrium again, there is a transition period, this time with inflow greater than outflow.

A temperature of 30°C, though, is hotter than our real Earth. How could we bring our EBM closer to a representation of reality? One way is to account for energy that leaves Earth's surface and proceeds directly into space. In the example above, we assumed that all the energy emitted by Earth's surface is absorbed, then re-emitted by greenhouse gases. But some portion goes right through the "atmospheric window" (look back at Figures 2.6 and 5.1). And it is not all radiation – some energy leaves Earth's surface through conduction and latent heat transfer. How would the equilibrium temperature calculated by our EBM change if we accounted for these flows? Our modeled Earth would be cooler, and closer to being realistic.

Treating the atmosphere as a single box is another large simplification. We know that the atmosphere has more molecules near the surface, where it is denser, and fewer molecules higher up, so we can get slightly closer to reality by dividing the atmosphere into two layers. With a "two-layer" model, the upper layer exchanges energy with both space and the lower layer, while the lower layer exchanges energy with both the upper layer and Earth's surface. Next, imagine a model that divides the atmosphere even more realistically into many layers, and examines energy flow that accounts for all the layers. Each layer has a different temperature – layers close to the surface are warmer than those at higher altitudes. In the real atmosphere, in addition to energy exchange among the layers via radiation, mixing via convection changes the temperature profile – thus, one might additionally approximate the effects of convection to represent Earth more realistically.

We could also get more specific about Earth's surface. We could assume a rocky planet or one completely covered in water, or we could partition the surface between oceans and land, using relative proportions similar to those here on Earth. Oceans and rocks have quite different heat capacities, which would change the trajectories of temperature in Figure 7.2 over time. And since the ocean water mixes (fairly quickly close to the surface, slower at greater depths), we also might include the rate of mixing and the volume of water that would have to be heated up or cooled down in the model.

Yet another step closer to modeling the real Earth would be to make calculations for different latitude bands, rather than averaging over the whole of the planet's surface. By doing so, we would gain information about the differences in energy balance at the tropics versus the poles. The tropics receive more energy from the Sun than they radiate into space, while the poles radiate more energy into space than they receive from the Sun. For the tropics not to overheat and for the poles to avoid a super deep freeze, there must be a net flow of energy from the tropics to the poles, via the atmosphere and oceans.

We have spent some time on EBMs to illustrate that it is relatively simple to build a climate model that can be useful for answering some climate-related questions, but a more realistic representation of Earth's climate requires more complex models.

7.2.2. Earth system models of intermediate complexity

The energy balance models we described above focus on overall energy flows with large-scale averaging. Earth system models of intermediate complexity (EMICs) are a step towards representing Earth's climate system in more detail. These models include some recognizable version of the distribution of continents and oceans on Earth, and typically include simplified representations of the large-scale circulation patterns in the atmosphere and oceans. In constructing EMICs, of which there is a huge variety,[10] scientists deliberately use a lot of parameterizations to avoid the computational cost of explicitly modeling many processes.

The relatively few calculations required for EMICs allow modelers to complete many runs fairly quickly, each using slightly different, but plausible, conditions. The ensemble results of these many runs help test the importance of different components of the climate system, and help answer questions such as: what are plausible ranges for the strength of feedbacks involving vegetation? are they amplifying or stabilizing? how much CO_2 might the oceans absorb under different conditions?

Another advantage of simplicity in EMICs is that they can model the climate over long periods – say, paleoclimate conditions far into the past, or projections thousands of years into the future. Ensemble approaches that combine many different model runs allow modelers to estimate the probabilities of what might happen in the future with higher confidence – for example, what is the probable range of climate response to the melting of the Greenland ice sheet? if we stabilized the rate of human CO_2 emissions today, how much additional warming might occur, and how fast?

Simplicity is also a disadvantage of EMICs. They do not yield detailed information about many processes in the climate system, nor do they achieve high spatial resolution. Despite their limitations, however, EMICs generally reproduce Earth's basic climate patterns – such as spatial patterns of temperature and precipitation – that agree broadly with observations and with more detailed models. They can also be used to identify particular intervals of time for which more detailed modeling might be illuminating.

7.2.3. General circulation models

General circulation models (GCMs) are more detailed representations of the Earth system. Their primary advantage is that they explicitly model more processes and rely on fewer parameterizations than do EMICs. GCMs are thus better tools for modeling dynamics in the atmosphere and oceans. They realistically simulate the general atmospheric circulation, with air rising in the tropics and descending in the subtropics. They realistically represent annual precipitation in sub-Saharan Africa and seasonal changes in surface wind speed. They reproduce equatorial Pacific sea surface temperature changes associated with El Niño cycles. They can simulate eddies in the oceans and storms in the atmosphere. They can be coupled with sea ice models, or ice sheet models, or vegetation models, to attempt to resolve those parts of the climate system explicitly.

With GCMs, modelers can investigate particular parts of the climate system in detail by turning things on or off. For example, modelers can add a lot of fresh water to the North Atlantic and see what happens to ocean circulation. They can increase atmospheric CO_2 at a specified rate, see where the carbon goes, and how Earth's temperature changes. They can test how changes in atmospheric circulation influence cloud cover. In short, GCMs are the best tools we have to run detailed virtual experiments with Earth's climate system.

A distinct disadvantage of GCMs, however, is that modeling more processes at higher spatial and temporal resolution means more equations to solve, and thus more computational time and expense. The higher the resolution modelers choose, the shorter the practical time the model can be run. Typical grid cells in a GCM might be 100–200 km on a side, with 20 vertical layers in the atmosphere and 20 more in the ocean – although grid cell sizes used in GCMs keep getting smaller with increasing computer power. As modelers move towards even higher spatial resolution – approaching a few kilometers square – the time step needed correspondingly decreases, and the length of time that is practical to model decreases to a decade or two.

Like EMICs, GCMs are used to estimate probabilities of future climate. They are, however, limited to shorter periods, typically running only a century or so into the future. Given Earth's radiation imbalance today and the projected imbalance for the coming century, these runs thus rarely show the system reaching an equilibrium state within the time period modeled.

7.2.4. Regional climate models

We are getting better at understanding the influence of humans on the global climate system. As the planet gets warmer, high latitudes will warm faster than the tropics. The Mediterranean will become drier, and the tropics will be wetter. But this is like saying you should carry an umbrella in Spain because it is going to rain in Vietnam. To plan for and adapt effectively to climate change, we need information about the future climate at much finer scales than general circulation models can provide. To decide whether to put up a dike, move some houses, switch crops, or buy insurance, we need data at scales of less than 100 km. One approach is to embed a finer-scale model of a particular area of interest into a larger-scale GCM. No region is isolated from the rest of the planet, so the GCM part of the model can keep track of what is going on globally, and exchange information with the finer-scale regional climate model (RCM).

First developed in the 1980s, RCMs have now been applied to almost all land surfaces on Earth and to most oceans.[11] RCMs take the data – on temperature, precipitation, and humidity, for instance – that GCMs provide about regions in short time slices of fifteen minutes. They then enhance these data with actual observations from, for example, satellites and weather stations, and they incorporate the effects of complex topography and coastlines, vegetation, inland bodies of water, and short-lived atmospheric particles. The modeler has to predict the next hour or so, then the RCM ingests more observations. Essentially, the work of the RCM is to compute a higher spatial resolution than is possible in a GCM. Rather than being "forced" by data on greenhouse gas concentrations, the RCM ingests the multimodel output of GCMs, paired with as much observational data as the modeler can obtain.

Regional climate models, however, are tricky. Modelers have to deal with the challenges of exchanging mass and energy at the regional border with the lower-resolution GCM. Often, a GCM drives an RCM, but the RCM does not return information to influence the GCM, although, in some cases, modelers have developed ways to deal with exchange in both directions.[12]

A major benefit of RCMs is that they are much better than GCMs at telling us something about extreme events. Floods, hurricanes, and heat waves are the result of relatively localized topography – which GCMs simply cannot capture. RCMs are more expensive than GCMs to run, however, and are best used with mountains of observational data. Unfortunately, such data do not exist for many parts of the world, and the research community is just beginning to apply RCMs to the task of adaptation. Given the value of robust data about regional climate change to local planners, engineers, and emergency response teams, for example, this suite of approaches is likely to evolve rapidly in the coming years.

7.2.5. Integrated assessment models

Integrated assessment models (IAMs) are multidisciplinary attempts to incorporate social and economic aspects into the modeling effort, with interactions and feedbacks between human activities and the rest of the Earth system. Economic activity, demographics, political decisions, and energy choices all influence carbon emissions. Surface temperature and precipitation influence resource availability, such as forest and agricultural products. Like the other types of models, IAMs incorporate key assumptions and parameterizations. Because IAMs explicitly incorporate social and economic information, they are used to develop plausible future emissions scenarios to feed into climate models. The merging of scientific, social, and economic factors produces a richer range of information to help people weigh the costs and benefits of action and inaction. IAMs thus aim to provide forecasts that are the most complete and useful for policy-makers: about what aspects of the full system are we the most uncertain? what policies might best help us to avoid catastrophic climate change?

One of the key economic parameters that influence the outcomes of IAMs is their discount rate – in other words, is it worth putting money, energy, and effort into taking action now, or would it be better to wait? Models that highly discount future values – that is, those that emphasize short-term gain – tend to support wait-and-see strategies. Models that do not discount the future so highly tend to support early action. One well-known example using IAMs, the *Stern Review on the Economics of Climate Change*,[13] concluded in 2006 that early mitigation would be both cheaper and more effective than waiting.

7.3. Certainties and uncertainties

No model can represent the real world perfectly, and none can forecast exactly what will happen. What models can do, however, is estimate the likelihoods of different outcomes or the relative importance of different system components. For example, what is the plausible range of climate responses if atmospheric CO_2 doubles? What is the likely magnitude of the water vapor feedback? Uncertainty about the answers arises from natural variability in the climate system, uncertainty in the models themselves and how they represent the system, and, for forecasts, uncertainty about future human societal decisions.

Climate change does not march linearly into the future. Some years will be warmer, some cooler, within a range of natural variability. No climate forecast can tell us what the temperature will be like on a specific day two years from now. Models can never tell us exactly when each El Niño of the future will occur. But models can provide information about the likely extremes for which we might want to prepare. We have more confidence in models that can reproduce the observed range of year-to-year natural variability.

There are some aspects of the climate system that we do not fully understand and cannot yet represent fully in models; thus, these aspects are often parameterized or left out. Clouds and aerosols currently are prime examples of climate aspects with wide ranges of possible behavior. Feedbacks involving sea ice, continental ice sheets, and ocean heat uptake are also less well constrained aspects of climate models. In the IPCC 2007 report, uncertainty about the melting behavior of ice sheets led to conservative (low) modeled estimates of future sea-level rise. And models tend to fail to reproduce the extremes in precipitation observed in the real world.[14] Scientists are working towards narrowing the ranges of possibility for these pieces of the puzzle. In contrast, we have excellent knowledge about the radiative behavior of CO_2 in the atmosphere; as an aspect with high certainty, CO_2's physical behavior is unlikely to surprise us. Models simulate the cooling effects of volcanic eruptions well, and the combined influence of the water vapor and lapse rate feedbacks is fairly well constrained.

Probably the most important source of uncertainty in forecasting the future is societal uncertainty: what will humans do? Models need to incorporate estimates of future changes in energy in the Earth system, but most of those changes will come from things that are largely controlled by human activity, such as the concentrations of greenhouse gases and aerosols in the atmosphere. How much fossil fuel will humans burn? How quickly? How will land use change and how quickly? What if we find a way to capture the CO_2 from fossil fuel burning and keep it out of the atmosphere? What if we find a way to suck CO_2 out of the atmosphere on a large scale, and store it somewhere? We explore some of these issues in greater detail in chapter 8, including the complexity and unpredictability of human behavior and the challenges associated with creating socio-economic scenarios.

Scientists are continually adding more sophisticated and realistic representations of more climate-related processes to models. Spatial and temporal resolution is constantly increasing. These changes can represent advances in our understanding, but we also encounter what Naomi Oreskes calls the "complexity paradox": "A complex model may be more realistic, yet, ironically, as we add more factors to a model, the certainty of its predictions may decrease even as our intuitive faith in the model increases."[15] This paradox is a useful reminder that models are not exact reproductions of reality, but neither are they inherently untrustworthy. Rather, they are crucial tools for learning that require continual checks against physical principles, real-world observations, and one another. As we learn more, we incorporate more into models, and since the additional pieces have their own ranges of possibility, the range of possibility for the model output often widens. Furthermore, we should be careful to identify the aspects of the climate system about which we are quite certain – such as the conclusion that humans are influencing the climate – and those about which we still have much to learn.

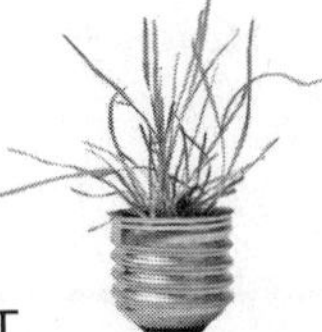

CHAPTER EIGHT

Future Climate: Emissions, Climatic Shifts, and What to Do about Them

- **8.1. Emissions scenarios**
  - 8.1.1. SRES scenario "families" and storylines
  - 8.1.2. Post-SRES and representative concentration pathways
- **8.2. The global climate in 2100**
  - 8.2.1. Temperature, precipitation, sea-level rise, and extreme weather
    - *8.2.1.1. Temperature*
    - *8.2.1.2. Precipitation*
    - *8.2.1.3. Sea-level rise*
    - *8.2.1.4. Extreme weather and abrupt changes*
  - 8.2.2. Uncertainty
- **8.3. Backcasting**
- **8.4. The scale of the challenge: Transforming emissions pathways**

MAIN POINTS:

- Future emissions scenarios are based on assumptions about population, economic growth, energy mix, and a host of socio-economic variables.
- Current global emissions are generally higher than the highest emissions scenarios created by the Intergovernmental Panel on Climate Change in 2000.
- Climate change projections for the twenty-first century include:
  - average warming of between 2.4°C and 6.4°C in a high-emissions scenario;
  - increasing intensity of precipitation events, including long stretches of drought in some places and floods in others;
  - rapid melting of the Greenland ice sheet , possibly crossing a threshold after which it will not recover; and
  - sea-level rise of from 0.2 m to as much as 1.0 m.
- The challenge of transformative change – reducing emissions by 80 to 90 percent by 2050 – requires early and ambitious mitigation strategies and adaptation to some climate impacts to which we are already committed.

The future is a foggy place, as dramatically different or as mundanely similar as one's imagination allows. Picture the world in 2100. Do cars fly? Is the planet bristling with megacities and towers that shoot hundreds of stories into the sky? Do we all speak the same language? What color is a sunset? What is the shape of coastlines?

Thinking about the future is a crucial dimension of the human imagination. It also lies at the core of our capacity to plan cities, respond to risks, and create better lives for our children than we experienced growing up. Even so, most of us tend to think of the future as a linear progression forward from the past. We struggle to conceive of surprises and radical diversions from familiar ways of living.

Day by day, however, our fossil-fuel-based energy system and our accelerating effect on the climate are creating an evermore uncertain future. To rein in this uncertainty, scientists around the globe have invested spectacular effort in developing models that illustrate the implications of our choices. As Chapter 7 showed, climate models help us to understand how the parts of the climate system interact with one another, but clouds, feedbacks, and other complex dimensions of the climate system inject considerable uncertainty into this process.

First, let us be clear about a few terms that tend to get jumbled when we talk about the future. When we speak about *projections* in climate science, we typically refer to model-derived estimates of the future climate if particular conditions occur. For instance, an ensemble of models could project that, given particular levels of greenhouse gases, the temperature might increase by, say, 2.2°C in the Pacific Northwest region of North America by 2080. Another projection, given different future conditions for greenhouse gases, could tell us that the temperature might increase by 6.2°C.

A *prediction*, or forecast, on the other hand, is the most likely projection, given what we currently know about input variables such as greenhouse gas emissions, Earth's reflectivity, and aerosols.

A *scenario* is a plausible, internally consistent "story" about the future. A scenario does not necessarily have a probability attached to it – in other words, there are scenarios that we think are very likely and others that are relatively unlikely. A scenario might use a projection as a building block, but then add elements to it, such as a starting point, or baseline, or information about policies, economic trajectories, and other factors. As with the "Great Transitions" initiative created by the Tellus Institute in Boston, we could tell an "eco-communalism" story, in which the preservation of natural systems and social solidarity are ensured through lower levels of consumerism, face-to-face democracy, and a local approach to resource distribution.[1]

In this chapter, we explore the implications of climate models through two lenses. The first is the particular *input* of greenhouse gas emissions in the climate models. Climate models are "forced" by such emissions, which, in turn, are the result of a deeply interwoven set of decisions about our energy system, urban design, agriculture, values, and policies. We explore the various emissions pathways that we might (or might not) choose, and we foreshadow the policy choices that we must make to change those pathways in the pursuit of sustainability.

The second implication of these climate models is the *output*: the effect of varying levels of greenhouse gas emissions on temperature, precipitation, sea-level rise, and extreme events such as hurricanes and heat waves. These are the direct impacts to which we must adapt as the climate changes, and which we might be able to mitigate through ambitious and creative greenhouse-gas-reduction strategies. We delve deeper into what these temperature and precipitation changes mean for both natural and human systems in Chapters 9 and 10.

8.1. Emissions scenarios

Consider the vast constellation of factors that contribute to our current level of greenhouse gas emissions. It is not as simple as thinking about just our sources of energy, although this is central. Our emissions are also determined by, among a myriad factors, how many people are consuming that energy, how far we drive to get to work, where our food comes from, what kinds of foods we eat and how they are grown, how efficient are our homes and cars, and what we demand for entertainment. At the roots of these behaviors are values – the status implied by owning a particular car, for instance, and the variety and novelty we expect in our day-to-day lives. The capacity to express these values through decisions is shaped by the availability and cost of options, social pressures, the politics behind policy choices, and even cleverly designed marketing campaigns.

In other words, emissions trajectories are the result of an array of socio-economic and political variables. As such, creating coherent, reasonable stories about the future requires the collaboration of economists, to understand the structure of taxes, trade, and consumer demand; political scientists, who investigate jurisdiction over emissions and collaboration between nations; engineers, who track and contribute to the emergence of new technologies; sociologists, who can tell us about norms and social movements; and a host of other specialists.

Because of the complexity of the human systems involved and the possibility of surprises, it is virtually impossible to predict future levels of emissions. Nonetheless, it is important for policy-makers to have some understanding of the long-term implications of a near-term policy choice: a tax on carbon, for instance, or a subsidy for renewable energy. These decisions might trigger an increase or decrease in greenhouse gas emissions, which will affect the scale of climate change we experience as well as the necessary adaptation or mitigation options.

Emissions scenarios weave together these variables to create a picture of the way our global, regional, or local greenhouse gas emissions might change. The output of these scenarios is then passed to climate modelers, who explore the implications for temperature, precipitation, sea-level rise, and extreme weather.

Over the years, teams of researchers have produced several "generations" of emissions scenarios that form the backbone of climate projections. The most influential set of scenarios are those in the Special Report on Emissions Scenarios (or SRES), a report published by the Intergovernmental Panel on Climate Change (IPCC) in 2000.[2] The scenarios in that report were used in the IPCC's Third and Fourth Assessment Reports, but the IPCC followed a new process in its Fifth Assessment Report.[3] Although we discuss both, we focus on the SRES, since these have informed most climate change scenarios to date.

The SRES scenarios are intended to be "baseline" scenarios, in that none assume any explicit greenhouse-gas-reduction policies. Rather, the scenarios are driven by a wide variety of factors – including demographic shifts, the structure of energy systems, and land-use changes – that fundamentally influence greenhouse gas sources and sinks, but that set the stage for any additional explicit efforts to reduce emissions. The SRES scenarios are neither predictions nor forecasts: no probability or likelihood is attached to any of them. Instead, the purpose of the scenarios is to paint different pictures of the future to allow us to grapple with the core drivers of greenhouse gas emissions and how these emissions affect our climate, and to decide what action we should take today.

The SRES scenarios are both qualitative and quantitative. In other words, they consist of narratives or storylines that describe what the future might look like, as well as of mathematical or technical models that illuminate numerical or measurable aspects of the economy, energy systems, and population growth. The scenarios cover the full range of radiatively important gases, including all major greenhouse gases as well as sulfur dioxide (which can have both heating and cooling effects on the climate).

8.1.1. SRES scenario "families" and storylines

Clearly, as we head towards 2100, there is an infinite number of possible futures. To make sense out of the chaos, the SRES process distills possible pathways down to only a handful of "iconic" scenarios, or "families," that share common drivers of emissions, economic trajectories, technological futures, and demographic shifts (see Figure 8.1). Policies affect these drivers, and range from transportation and land-use policies to social policies that address inequality or population growth. A narrative

Figure 8.1. Trajectories of Basic Drivers of Greenhouse Gas Emissions in the SRES Scenarios to 2100

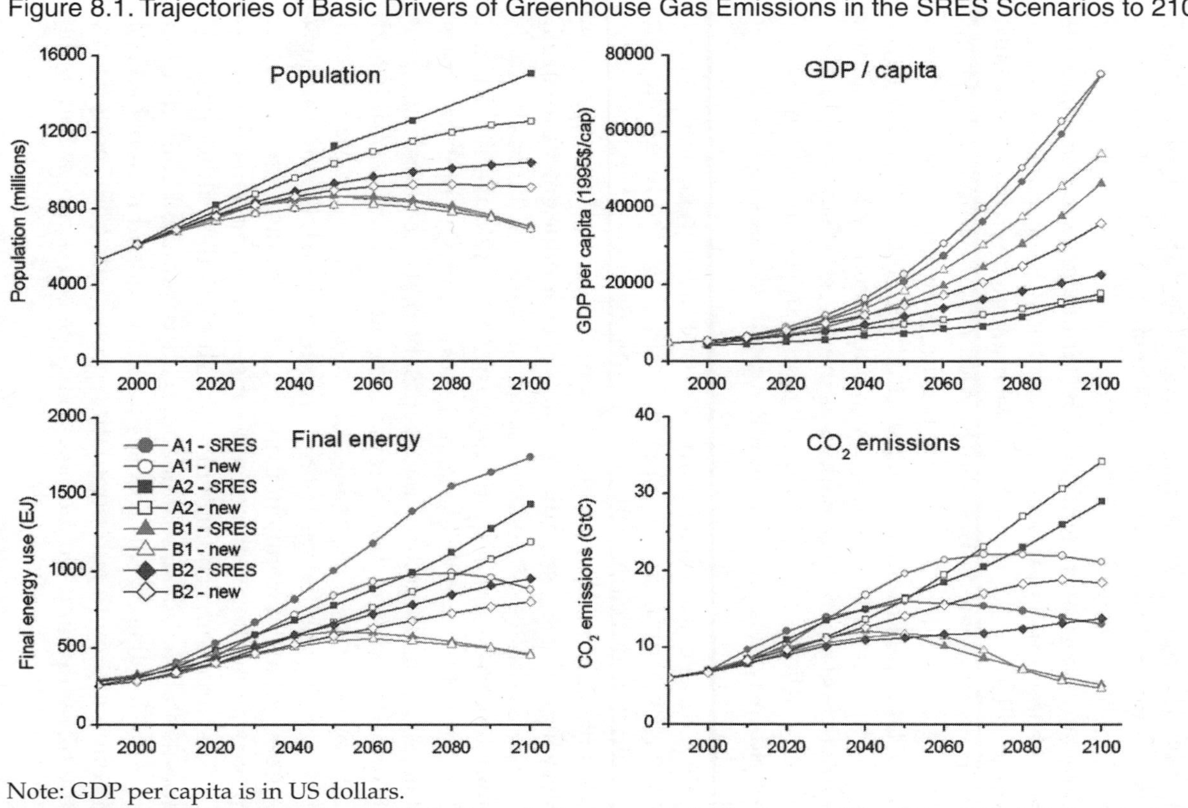

Note: GDP per capita is in US dollars.

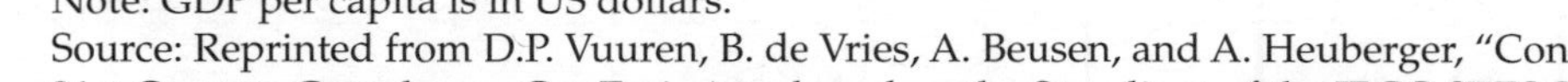
Source: Reprinted from D.P. Vuuren, B. de Vries, A. Beusen, and A. Heuberger, "Conditional Probabilistic Estimates of 21st Century Greenhouse Gas Emissions based on the Storylines of the IPCC-SRES Scenarios," *Global Environmental Change* 18, no. 4 (2008): 635–54. © 2008, with permission from Elsevier.

describes each of these families, weaving them together in a way that is logical and useful. Interestingly, the scenario-writing team decided that, due to the complexity and uncertainty described above, they would identify neither a "most likely" nor a "business as usual" scenario.

Roughly speaking, the "A" family of scenarios focuses more on economic growth, while the "B" family focuses more on environmental protection. The first family of SRES scenarios, unhelpfully called "A1," describes a technologically driven, socially and culturally globalized world in which levels of development (and thus fertility) converge to produce low population growth. This family contains scenarios that are fossil fuel intensive and those that are less so. It is characterized by high economic growth and a variety of pathways the energy system might take. An important variant of the A1 family of scenarios is A1FI – a world in which rapid economic growth is sustained by heavy reliance on fossil fuels and a very high emissions trajectory.

A2, the second SRES family, describes a heterogeneous world that is less uniformly globalized and thus produces varying patterns of development and fertility. Because of the focus on regional self-reliance and local identities, population growth is very high, but accompanied by slower economic growth overall.

The B1 family of scenarios brings in stronger themes of environmental sustainability than either A1 or A2. Like A1, B1 is a "convergent," or more globalized, world, so population growth is low. A crucial difference, however, is the path the global economy takes: B1 scenarios show a rapid transition towards services and information-based economies that consume fewer materials and have great potential for energy efficiency. In B1, global solutions are sought to environmental and social problems, rather than the more fragmented, regionalized approach assumed in A2.

B2 shares some characteristics with A2 in that, although environmental and social sustainability are paramount, the approach to reaching these goals focuses on the local and regional levels. Economic growth and population growth are both moderate in the B2 family, and technological change is less rapid than in the B1 or A1 families.

But what do all these projections mean? This would vary, of course, from country to country and even from city to city, but there are some commonalities. In a B1 world, for instance, you might be more likely to have an information- or communications-related job, rather than working in the manufacturing or resources sectors. You might take high-speed transit to work, but still eat foods from around the world that are transported on airplanes that consume biofuels. In other words, in this scenario, global trade and technological innovation are still crucial, and have contributed to highly efficient resources use and a services or information-based economy.

In an A2 world, strength comes through self-reliance. This might mean a muted conversation about international aid, or an aversion to jobs with companies owned by foreigners. More food is produced locally, but perhaps not sustainably. You might see stories on the news about your hometown, which has gone bankrupt and has difficulty providing basic services to its citizens, while other towns are doing just fine.

Clearly, there are pros and cons associated with each of these worlds, depending on one's own views, but a central difference between the worlds (for our purposes) is their effect on carbon emissions over time.

Critiques of the SRES scenarios take a couple of forms. One is related to the underlying data: the SRES scenarios use market exchange rates for converting national currencies into US dollars, rather than purchasing power parity (PPP) exchange rates (an assessment of the relative cost of goods and services in an economy).[4] Per capita income tends to be higher when measured using PPP, and since the SRES scenarios assume economic convergence between developing and developed countries, using PPP instead of market exchange rates shows less economic growth on the part of developing countries – and thus fewer emissions.

Another criticism of the SRES scenarios pertains to the population growth assumptions upon which they are built.[5] If families are smaller and children are born later in women's lives, population projections typically will be lower. Where an exploding population was once considered the bane of environmental sustainability, we now see that, in fact, per capita consumption of resources is a far greater culprit. Furthermore, the planet's population is "exploding" a little less quickly than we once thought, due in large part to better access to family planning, more diverse economic opportunities, and a suite of other development-related factors. Inappropriately high population estimates in the SRES scenarios mean that emissions projections might be higher in some cases than are now thought likely. As a result, researchers have been advised to take this into account when employing the SRES scenarios.

Despite these criticisms, however, the broad assessment of the scientific community is that the SRES scenarios were robust enough to play a central role in climate change projections for more than a decade.

8.1.2. Post-SRES and representative concentration pathways

The world of emissions scenarios changed rather dramatically with the IPCC's Fifth Assessment Report, released in 2013 and 2014. Rather than creating a new set of SRES-like scenarios, as the IPCC did in 2000, the IPCC has brought together scenarios created by the broader research community. The process itself is changing as well: for the SRES scenarios, socio-economic scenarios were first created – telling stories about population, economic growth, and technological change, for instance – which then led to quantitative and qualitative modeling of greenhouse-gas-emissions trajectories. These emissions figures were then fed into climate models to determine what the effects might be on climate change and, ultimately, what response options might be possible.

This time around, the scenarios *start* with "representative concentration pathways" (RCPs) that reach specified climate conditions by 2100. These conditions are defined by changes in energy in the climate system by 2100. For example, RCP8.5 suggests that, by 2100, there will be an additional 8.5 W/m^2 in the Earth system compared with pre-industrial times – the equivalent of having 1,370 ppm CO_2 in the atmosphere, compared with a pre-industrial concentration of 280 ppm.[6] Each of four RCPs has associated trajectories for greenhouse gas concentrations and other factors, such as aerosols, that influence energy fluxes, ranging from RCP8.5 – the high-emissions pathway – to RCP 2.6PD, in which emissions peak, then decline, before 2100. The RCPs, with their associated emissions trajectories, can be used in parallel by both climate modelers who are investigating the effects of these trajectories and integrated assessment modelers who are working backward to identify the socio-economic conditions that give rise to the trajectories. Put more simply, what was once a step-by-step, time-consuming process of *a* leading to *b* leading to *c* can now be done simultaneously. One effect of this approach is that scientists can produce scenarios more quickly – a huge boon given the pace of change and rapid emergence of new technologies and policies. It also facilitates quick integration of possible future human choices – for example, aggressive mitigation efforts that are not yet happening.

The Representative Concentration Pathway approach is just now emerging, and the implications for projecting future climate and how to incorporate this approach into policy-making are not yet clear.[7]

8.2. The global climate in 2100

8.2.1. Temperature, precipitation, sea-level rise, and extreme weather

Greenhouse-gas-emissions levels resulting from socio-economic assumptions are fed into climate models, which then yield projections that describe how climatic variables would be "forced" given different levels of emissions. In the case of the IPCC reports, the changes in variables that are described are actually the amalgam of the results of many models, which helps to iron out differences between models and yield a more reliable projection.

The effects of increasing emissions of greenhouses gases can be categorized as either "first order" – such as temperature, precipitation, and sea-level rise – or "second order" – such as lower agricultural production in some regions due to changing temperature and precipitation, or the degradation of coral reefs due to warmer ocean temperatures and acidification. Rather that meticulously presenting the projections made under both high- and low-emissions assumptions, we focus mostly on high-emissions worlds in the sections that follow. This is for good reason: the evidence shows that actual global emissions continue to climb steeply – in fact, we are emitting even more than in some of the worst-case emissions scenarios. This means that we are potentially facing even more extreme changes to the climatic system than we thought, and greater impacts that will result from these changes.

8.2.1.1. Temperature

Generally speaking, climate scientists are very certain that, if emissions levels continue at their current rate, or increase, we are likely to see accelerated warming of the planet. Early in this century – that is, to 2030 – the global average temperatures is likely to be approximately 0.65°C warmer than during the 1980–99 period.[8] This warming is likely to occur essentially regardless of which emissions pathway we follow, because of time lag in the climate system and the warming to which we are already committed. By mid-century, however, the scenario we follow will begin to make a significant difference, and the uncertainty in these mid- and late-century projections also increases (see Plate 2). By 2100, for instance, the planet might experience anywhere from 2.4°C to 6.4°C of warming in a high-emissions – that is, high-population-growth, fossil-fuel-intensive – scenario. The range would be even wider, of course, if we look at lower-emissions scenarios. The SRES A2 storyline suggests that we could reach around 3.4°C of warming by 2100, but only 1.8°C in the B1 scenario, which includes a stronger emphasis on environmental sustainability and global solutions. This might not seem like much, given that the temperature in any given day can swing by three or four times this amount in many parts of the world. These temperature swings, however, are simply regional temperature fluctuations, not the number we get when we average thousands of measurements every day, year after year. It takes a lot to nudge global temperature averages up, but when they do go up, it is very likely that heat waves will become more intense, last longer, and become more frequent, which will put significant stress on ecosystems (see Chapter 9) and on vulnerable humans.

The newer RCP approach echoes the findings of the SRES. Because global emissions crept steadily upward by approximately 1.9 percent per year in the 1980s, 1.0 percent per year in the 1990s, and have risen more than 3 percent per year since 2000, we are currently looking at the SRES's higher-emissions scenarios.[9] As you can imagine, global average temperature increases have been at the top end of the scenarios as well, and scientists have suggested that the threshold beyond which we enter a world of "dangerous anthropogenic climate change" is 2°C of warming.[10] There is, of course, debate over the meaning of this phrase, with some scholars arguing that "dangerous" can be defined in terms of physical vulnerabilities, such as the breakdown of ocean circulation, or social vulnerabilities, such as the breakdown of social order resulting from a massive influx of environmental refugees.[11] If we are to avoid reaching this threshold, we need to achieve significant near-term – beginning before 2020 – emissions reductions. Furthermore, we need to reduce emissions by approximately 3 percent per year, rather than increasing them by about this much, as we are now – so that they peak before 2020.[12]

The main lessons here are threefold. First, it is clear that our current levels of emissions are putting us on track to a much warmer planet than we once projected. Second, if we are to stabilize emissions early enough to avoid warming in excess of 2°C, then we need to put mitigation policies in place immediately. Third, uncertainty in our temperature projections increases later in the century, opening up the possibility of feedbacks and surprises that will make the future a foggy place indeed.

8.2.1.2. Precipitation

In addition to an increase in the planet's average temperature – with more warming at higher latitudes and over continents – scientists also expect a wide variety of changes to precipitation. This means that a one-size-fits-all response strategy will not work, and that we need more information about which crops are grown, what freshwater resources are available, and how to deal with the potential for more intense instances of floods or droughts.

Despite this variety of changes, we can make one general statement: warming will lead to an *intensification* of the hydrological cycle. What does this mean? In all the scenarios in the IPCC's Fourth Assessment Report, average global precipitation increases, especially where precipitation is already high, such as in the tropics and Asian monsoon region.[13]

The crucial element of changes in precipitation, however, is not simply the total annual amount, but *when* the precipitation comes – does it fall in occasional great deluges or in smaller sprinkles throughout the year? The answer will have severe implications for agriculture, ecosystems, and freshwater resources. Some regions – such as between 10° of latitude north and south of the equator – will experience a significant increase in overall precipitation in a warming world,[14] which will contribute to higher levels of soil moisture and runoff, which, in turn, could mean flash floods and increasingly unstable slopes, leading to mudslides and property damage. Other regions, such as the Mediterranean, the Caribbean, and the subtropical western coasts of each continent, likely will experience a significant *decrease* in precipitation. Widespread decreases in mid-latitude summer precipitation are also expected; indeed, we are already getting a taste of this: in early 2013, for instance, approximately 70 percent of the contiguous United States and significant portions of Mexico and central Canada were experiencing moderate-to-severe drought conditions.[15] These changes ultimately will also affect river runoff: river flows could be higher in high latitudes (where it will get wetter), but much lower throughout the Middle East, Europe, and Central America.

Consider what this means for agriculture, drinking water, and sanitation in some of the world's most populous regions.

To sum up the precipitation scenarios: the patterns are not simple. Some regions will experience flooding while others will see significant drought. Some regions will experience *both* drought and floods, amplifying potential damages and costs. This suggests the need for robust regional climate models and a wide variety of climate change adaptation measures.

8.2.1.3. Sea-level rise

Most of the sea-level rise that we observe occurs for two reasons: either the water actually takes up more space as it warms – this is called thermal expansion – or there is more of it due to the melting of land-based glaciers and ice sheets. Both are the result of a warming climate, and are affecting vulnerable coastal ecosystems and human settlements. In 2007, the IPCC projected that sea-level rise during the twenty-first century due to thermal expansion would be between 14 and 35 cm in a high-emissions (A2) scenario. To this projection, the IPCC added estimates of flows of fresh water melting from land-based glaciers and the western Antarctic and Greenland ice sheets, which could boost the global average sea-level rise to between 23 and 51 cm over the coming century.

Since those 2007 projections, however, deeper concerns have emerged about the stability of the both the western Antarctic and Greenland ice sheets. In the 2007 IPCC report, scientists suggested that, although the sudden loss of either of these ice sheets could trigger a sudden sea-level rise of between 5 and 7 m, models at that time showed this occurrence to be unlikely in the twenty-first century.[16] Since then, new research has suggests that the IPCC's estimates were too modest, for two reasons. First, melting sea ice – which often gets overlooked because it is already in the water and thus does not add to the *volume* of sea water – causes the ocean to become darker (recall our earlier discussion of albedo) and speeding up warming even further. Second, the melting of this ice pumps cooler fresh water out and exchanges it for warmer, saltier water, setting up a giant oceanic heat pump that also accelerates warming.[17] Warming then speeds up the rate at which the ice sheets melt, raising sea level. Some scientists argue that this puts us on track for at least a meter of sea-level rise in this century, rather than the 0.2–0.5 m the IPCC projected in 2007.[18] Indeed, recent observations back this up: sea levels are rising 60 percent faster than the IPCC projected,[19] sparking new concerns about the vulnerability of coastal cities and settlements.

8.2.1.4. Extreme weather and abrupt changes

One of the more challenging aspects of a changing climate is that, as average temperatures go up, changes in other climatic variables are not steady or predictable. Projections suggest that, in a high-emissions scenario (such as SRES A1F1), we will see increased severity and frequency of both extreme drought and extreme floods. This is because precipitation will be clustered into periodic, intense events, with long dry spells in between. This is likely to lead to an increase in summer flooding in Europe,[20] greater summer flooding in the Asian monsoon region,[21] and more intense, longer-lasting heat waves.[22] The type of heat wave Europe experienced in 2003, in which more than 15,000 people died in France alone,[23] could become more common in a warming climate.

A warmer planet would mean that more energy is injected into the climatic system, which would affect wind speed and, as we saw above, precipitation intensity – two key ingredients of hurricanes or tropical cyclones. Projections for this century indicate that hurricanes are likely to become more severe, with greater wind speeds and more intense precipitation. Hurricane Sandy, which battered the east coast of the United States and the Caribbean in October 2012, gave a taste of the type of storm that might become more frequent in a changing climate. Although no single storm can be linked conclusively to climate change, these examples can help us to better understand the ways in which communities are vulnerable, and to prepare proactively for future events. If warming is putting more energy into the system, it is fair to say that the "fingerprint" of climate change is in every storm even if we cannot say what the storm would have been like in the absence of warmer average temperatures. Some scientists project that the overall number of storms actually might decrease, but this masks a complex pattern: projections indicate that the number of *weaker* storms could diminish, but the number of *extreme* storms could increase.[24]

In addition to the more familiar extreme events such as hurricanes, scientists are also exploring the potential for abrupt or extreme changes to ocean circulation and the mass of land-based ice – that is, rapid, nonlinear shifts that are outside the natural range of variability. Examples are the collapse of the Gulf Stream, which delivers warm water to the North Atlantic or the sudden, dramatic loss of the Greenland ice sheet. The Gulf Stream circulation is shaped in part by differences in the Atlantic Ocean's density – since salt water is denser than fresh water – which, in turn, generate overturning that pushes forward a "conveyor belt" of heat. Models universally predict a weakening of this current as water freshens due to the melting Greenland ice sheet, although none of the models run for the IPCC's Fourth Assessment Report (in 2007) show the Gulf Stream completely shutting down.[25] At the same time, observations suggest that the Greenland ice sheet is melting much more quickly than anticipated,[26] and that, sometime this century, it could reach a threshold from which it cannot recover.[27]

8.2.2. Uncertainty

Despite the ever-growing body of research on climate change, the complex and interrelated processes that drive climate change continue to be characterized by significant uncertainty. Policy-makers and practitioners – such as municipal flood management engineers and urban planners – must make very specific decisions based on the output of climate models, but how can they account for uncertainty while simultaneously pushing forward proactive responses to the risks?

A great unknown is the extent to which human actions might alter the climate over the decades and centuries to come; instead, we have widely varying assumptions that alter the type, rate, and extent of projected effects.[28] About some things, we are reasonably certain. For example, we have some idea of how much conventional oil remains, although even this idea shifts as market forces make formerly uneconomical deposits viable and as new reserves are occasionally discovered. Whether we choose to exploit these reserves, however, is the product of complex social systems, making long-term predictions virtually impossible. The parties that are voted into power in key nations, the conflicts that are ignited, and the cumulative ripple effects of policy decisions are among the many uncertainties that cannot be "solved" simply through acquiring more or better data. Given the pace of technological change and the chaotic nature of geopolitics, the past is not always a good guide to the future. The assessment of changes in components of the climatic system – including predictions of the effects of greenhouse gas concentrations on changes in climatic variables, radiative forcing, climate response, and impact sensitivity – remain highly uncertain as well.[29]

In light of such uncertainties, current attempts to better understand the implications of a changing climate are based on assessing the outlook for future emissions (and emission reductions) of greenhouse gases and aerosols, the resulting changes in climatic variables and their effects on ecosystems and society, and the extent and effectiveness of actions to adapt to or ameliorate the effects.[30]

A traditional view of uncertainty, embodied in much climate change research and commentary, holds that a lack of understanding of human responses to climate change is best addressed by more and better analysis.[31] In other words, more research will lead to better understanding and thus to better predictions. An alternative view, however, argues that, because of the existence of intentionality, the actions of human systems are exceedingly complex and inherently impossible to predict.[32] Accounting for the human dimensions of change, therefore, requires specific attention to human choice, and indicates that certain common practices or assumptions, such as attempting to determine the likelihood of certain effects, might not be useful concepts. If we are unable to attach probabilities to scenarios depicting socio-economic, political, and cultural systems,[33] then an additional, possibly greater, form of uncertainty exists in all climate change scenarios. Such arguments give rise to approaches to studying the future of human systems based upon various forms of scenario analysis, including backcasting.[34] We explore this form of scenario creation in section 8.3.

8.3. Backcasting

Sometimes, what is most valuable is not where we are *most likely* to be but where we *want* to be. Should we not make decisions about what we value, the future we want to see, and then explore what it takes to get there?

The process of building and using scenarios typically starts from a particular set of assumptions – about population, the structure of the energy system, the pace and direction of technological change – then moves into the future to suggest where we might end up. For example, if we do not adopt solar power quickly enough, can we stabilize greenhouse gas emissions through wind and bioenergy? If population rises rapidly, will greenhouse gas emissions follow suit?

A very different approach to the making of scenarios is backcasting. Once we determine what kind of future we *want*, rather than what future is most *likely*, the task then becomes one of working backward from the desired end point to determine how to achieve it.[35] Originally developed in the energy field to allow identification and analysis of energy futures that were invisible when the more traditional predictive forecasting approach was adopted,[36] backcasting has undergone a revival since 1990. The approach has been extended to include a more general analysis of alternative desired futures, especially those focused on sustainability, at scales ranging from the national to the local.[37] Early backcasting studies tended to focus on scenarios researchers created to explore the achievement of externally defined targets, but a prominent characteristic of some newer backcasting studies has been their participatory nature, in which the future scenarios themselves are the product of deep and ongoing engagement with a range of stakeholders, including the public, policymakers, and practitioners.[38] Here, the focus of the analysis shifts from backcasting from external targets to backcasting from the participants' preferred futures.

These two dimensions of backcasting – that it can be participatory and that it makes explicit the values embedded in our choices about the future – differentiate it from other scenario processes. After all, sustainability is largely a value proposition: we make choices about how much climate risk we are willing to accept, what our notion of progress is, and what role the natural environment plays in our lives. Backcasting is being used increasingly in community planning processes, and is opening up a different way of thinking about the future.

8.4. The scale of the challenge: Transforming emissions pathways

Globally, the news is not good: our total greenhouse gas emissions continue to climb, putting us on track for increasing temperatures, changing precipitation patterns, rising sea levels, and increasing severity and frequency of extreme weather.

But the global average masks very different local realities. Germany is leading all countries in its implementation of solar power on a massive scale; the United Kingdom is on track to meet its emissions-reduction commitments under the Kyoto Protocol; Masdar City, in the United Arab Emirates, is an experiment in carbon-neutral, zero-waste community building. From Vancouver to Amsterdam, active transportation is taking the place of the traditional commute, and values are shifting in some urban areas in favor of public transit, compact, complete communities, and inspiring approaches to sustainability.

These examples of successful action on climate change bring our attention squarely to the issue of what world we would *like* to see in 2100, and how we should get there. If emissions must peak by 2020 and decline by 80–90 percent by 2050 to avoid the worst impacts of climate change, what must we do to make this happen? This is an open question, but it requires a careful examination of our energy policies, building codes, land-use and transportation plans, and a host of other tools. Switching off the lights – especially in regions where electricity comes from hydro power – is not sufficient. Transformative change requires not just technological advances, but also a shift in values, habits, social norms, and policies at multiple levels of government. We explore this in more detail in Chapter 11.

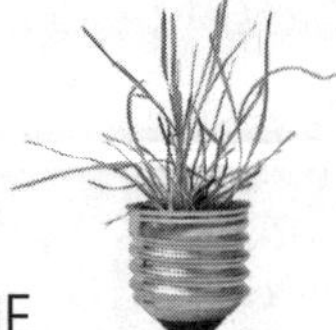

CHAPTER NINE

Impacts of Climate Change on Natural Systems

MAIN POINTS:

- The vulnerability of natural systems to climate change impacts is a function of a host of factors, including exposure, sensitivity, timing, and the probability of impacts.
- We are already seeing the impacts of climate change in terrestrial and aquatic ecosystems.
- Currently observed impacts include changes in the timing of lifecycle events, shifts in the range and morphology of species, coral reef bleaching, ocean acidification, and a rise in the number of endangered species.
- Enhancing biodiversity is a key strategy that will contribute to the resilience of ecosystems.
- Global, national, and subnational policies have emerged to conserve biodiversity, including the Convention on Biological Diversity and the growing momentum behind ecosystem-based approaches to adaptation and mitigation.

Over the past decade, climate change scholars and activists have struggled to put a human face on climate change. We are confronted with images of drought, preventing millions from accessing clean water, of floods and the livelihoods lost, and of food shortages afflicting already-battered populations. At the root of these calamities are the impacts of climate change on natural systems. Whether you feel that these impacts are important only insofar as they affect human well-being or that these systems possess inherent value that must be preserved, climate change poses a real threat to the stable functioning of our planet's ecosystems.

In this chapter, we consider those climate change impacts that have already begun to be observed around the globe, with a particular focus on those on terrestrial biological systems (including plant and animal species), coastal systems, and aquatic systems (oceans, lakes, and rivers). These impacts are directly related to the capacity of the planet to respond to shifting climatic patterns and to provide for the needs of humans.[1]

When we speak of climate change impacts, we are referring to those phenomena that occur as a result of rising temperatures and shifting precipitation patterns, and they can be either positive or negative. Rising sea levels, for instance, are a second-order impact of climate change, which occurs primarily as a result of first-order impact – namely, melting land-based glaciers and the thermal expansion of water. Rising sea levels, in turn, might lead to flooding, erosion damage, and the altered distribution of ecosystems – what we call third-order impacts. In other words, climate change impacts are the "damage report" from the mass combustion of fossil fuels, and they present monumental challenges to both human and natural systems. But how vulnerable are natural systems, in particular, to the impacts that comprise this "damage report?"

Vulnerability to climate change represents the convergence of exposure to impacts and sensitivity to them. In other words, we generally would not consider a system highly vulnerable, even if it is extremely sensitive, if it is also highly unlikely to experience the impacts of climate change.

The following criteria help to identify key vulnerabilities to climate change impacts:

- the magnitude of impacts;
- the timing of impacts;
- the persistence and reversibility of impacts;
- the likelihood (or probability) of impacts;
- adaptive capacity;
- the importance of at-risk systems; and
- the distribution of impacts.[2]

When we apply these criteria to human and natural systems, we see that coastal systems, food, the supply of fresh water, and infrastructure are particularly vulnerable to the effects of climate change as we move into an uncertain future. These criteria are largely characteristics of natural systems or physical infrastructure, but in Chapter 10, we explore other dimensions of vulnerability and adaptive capacity as they pertain to the impacts of climate change on human systems.

9.1. Observed impacts

9.1.1. Impacts on land

After decades (some might say centuries) of studying ecosystems, it has become clear that plants and animals function best within a particular climatic range. Water availability, ambient temperature, and the balance of nutrients all dictate the success of one type of ecosystem or species over another, and have yielded the current distribution of ecosystems that is present today. Of course, ecosystems are not stable, static phenomena – they are constantly shifting, evolving, and being interrupted by fire, pests, predators, and human activity. Although disturbances can enhance species and genetic diversity, extremes reduce them.

Ecosystems change through the natural process of succession. In a typical forest ecosystem, for instance, this process might begin with the colonization of newly exposed or recently disturbed land by annual plants, followed by perennials, shrubs, softwood trees, and finally hardwoods. As ecosystems mature through these successional stages, they become more diverse and complex. This also might mean that they provide a greater range and abundance of *ecosystem services* to humans, including the provision of food, clean water, and medicines, the control or moderation of climate, the support of nutrient cycles and crop pollination, and important cultural or recreational benefits.

One of the most compelling drivers of the conservation of biodiversity – the quantity of species, genetic material, and ecosystems around the globe – is actually what we *do not* know about ecosystems. Current best guesses suggest that there are around 8.7 million eukaryotic species – that is, plants and animals, but not bacteria or viruses – on earth. Most of these are a complete mystery to the scientific community: scientists estimate that we have discovered only around 14 percent of land species and 10 percent of marine species.[3] Why does this matter? Consider that about 25 percent of the prescription drugs used in the United States are derived from plants. Lose a little biodiversity, and you potentially lose a shot at the next miracle drug. This loss is also irreversible: we are losing species we have never even encountered before, so there is no way to recapture or retain the genetic resources associated with those species. The ripple effects that cascade out from the loss of a particular species can be dramatic, as well. Each species in an ecosystem has a role to play, so the loss of a single species might affect the healthy functioning of the ecosystem as a whole, placing other key species at risk.

On average, ecosystems exist within a particular range of climatic conditions, and if these conditions change dramatically (or rapidly), plants and animals respond in several ways. Timing of key events such as the blooming of flowers or mass migrations of animals might shift to earlier or later in the season. Populations might shift to a new location that provides the conditions under which they function best. Under extreme stress, species might become extinct altogether.

9.1.1.1. The changing timing of events, migration of species, and altered morphology

In a warming climate, species often respond by moving towards the poles or to higher elevations. Particular genetic varieties might find greater success, while others fall away, yielding changes in shape (morphology) or reproductive habits. Finally, if suitable conditions cannot be found, plant and animal species might simply become endangered or extinct. We find this with species that are particularly constrained geographically, or are unable to adapt.

Earth is currently in the middle of a human-caused mass extinction. Rather than a process that is tracked over time, extinction is more like a day that can be marked on a calendar. Once a species goes extinct, all the diversity that it represented is lost forever. Mass extinctions and the more frequent individual species extinctions mean that most of the species that have ever existed are now extinct through natural processes. Ecosystems might be destroyed by severe disturbances, but they do not really go extinct unless the species that make them up are lost. Today, habitat degradation, pollution, overfishing, and hunting are straining ecosystems and species to the breaking point. Climate change, however, puts a *systemwide* strain on ecosystems, rather than the more familiar pressures on individuals or groups of species.

In response to higher temperature, the leaves of some tree species in North America are unfolding progressively earlier in the spring. Similarly, tree lines – the boundary between habitable and uninhabitable environments for trees – are shifting to both higher elevations and higher latitudes. But why should we care where trees are, as long as they still exist *somewhere*? Economically valuable tree species might be moving into new territories where the forestry industry's infrastructure – logging roads, pulp and paper mills, and labor – does not exist. This gets even more complicated if tree species can no longer survive in a particular jurisdiction but flourish elsewhere. This is one of the complex dimensions of climate change impacts: some groups will win, while others lose. Those forests also contain a particular mix of plant and animal species that might be economically and culturally important to nearby communities, some of which are playing only a very small role in the climate change that is causing these shifts.

In Europe, as in North America, bird species appear to be particularly affected by the warming that has already occurred. In fact, birds are often viewed as "indicator species" – the proverbial canaries in the coalmine – of climate change impacts. Some species, for instance, are laying eggs earlier in the season, and several species no longer migrate out of Europe. When the timing of these key events shifts, bird populations might be pushed out of sync with the lifecycles of insects and plants upon which they depend.

Powerful evidence for warming in the United Kingdom is provided by the significant northward shift in 329 species of birds.[4] These species are taking advantage of an extension of their range into previously cool regions, while the southern regions of their habitats might soon contract as temperatures rise to levels outside their coping range. Species that prefer a narrow range of environmental conditions are expected to suffer the most, as well as those that call islands, mountains, or Arctic or Antarctic habitats home.[5] Sea-level rise will continue to affect the intertidal zone that provides crucial habitat to thousands of bird species, and the development pressure on mangroves and marshes is creating an additional source of stress. Consider the benefits of bird species as predators and prey, carriers of pollen and seeds over vast distances, managers of insect populations, and the objects of affection for birdwatchers the world over.

Amphibians are another class of animals that is particularly sensitive to changing climate, and that have been devastated by warmer temperatures.[6] Several direct and indirect impacts of climate change, rather than a single cause, are responsible for the observed decline in amphibians. Environmental contamination, loss of habitat – due to both development pressure and the shifting ranges that result from altered temperature and precipitation regimes – introduced exotic species that compete with existing species for resources, and even increasing ultraviolet-B radiation are all having a mutually reinforcing (and deeply negative) effect on amphibians.[7] One impact in particular seems to be causing mass mortalities in amphibian populations: scientists have discovered that, in some regions, warmer weather is leading to the explosion of a chytrid fungus, which has been implicated in the decimation of 67 percent of the 110 species of frogs in the genus *Atelopus*.[8]

These phenomena provide some idea of the complexity, and irreversibility, of climate change impacts. In some wetter, warmer regions, pathogens will flourish; in others, drought will prevent the spread of pathogens that have controlled populations in the past.

A warming climate and changing precipitation patterns also affect terrestrial ecosystems that are of direct value to humans. Longer growing seasons and the fertilizing effect of CO_2 in the atmosphere actually have caused an increase in net primary production, or vegetation growth, in both the northern hemisphere and the tropics. Plate 3 shows estimated increases in plant and tree growth during the 1982–99 period.[9] In some places, though, this increase in plant growth has been offset by the effect of pests such as the bark beetle, which thrives in warmer temperatures and has been chewing its way through forests in the United States and Canada. This raises a particularly important issue: changes to the climatic system can trigger amplifying feedbacks that further enhance the effect. The damage done by bark or pine beetles is an example: the larvae of these pests are better able to survive through warmer winters, causing an explosion in their populations. The beetles leave huge swathes of dead trees in their wake, which ultimately decompose and return carbon to the atmosphere. Of course, these trees are also no longer absorbing carbon, which further enhances the carbon burden on the atmosphere. By 2020, the cumulative effect of just this most recent pine beetle outbreak in western Canada alone is estimated to be 990 megatonnes of carbon,[10] equivalent to about five years of emissions from the Canadian transport sector.

On the agriculture front, some crops are expanding their range in a changing climate, but others are not doing so well. Maize and soybean crops, for example, are showing increased yields in some regions, but hay yields in the United Kingdom and rice yields in the Philippines have gone down. The impacts of climate change on agriculture are deeply contingent on water availability and management practices and on the exposure of arable land to climatic extremes. In China, for instance, which is home to 22 percent of the world's population but only 7 percent of the world's arable land, glaciers have retreated and the frequency of heat waves has increased.[11] Dry regions in the north of the country have become even drier, and wet regions in the south have become wetter. Meanwhile, management practices have led to lower yields of rain-fed crops such as wheat and higher yields for irrigated crops such as rice.[12] If rice yields are going down in the Philippines but up in China, what lessons can we learn overall? If there is one message here, it is that climate is extremely variable from place to place, and regional climate models and local adaptation are necessary to understand the specific impacts that any given community might experience.

While evidence is accumulating for the impact of climate change on terrestrial systems around the globe, in no place are these impacts more evident than in the Arctic and Antarctic. Higher latitudes tend to warm to a much greater extent than mid- and low latitudes. The Arctic, for instance, has warmed by approximately 2°C per decade over the past thirty years.[13] As a result, tundra in Alaska and the broader Arctic has been gradually giving way to shrubs and other vegetation.[14] One might think that the transformation of a rather barren-looking landscape into one that is lush and growing is a good thing: but fragile and valuable ecosystems are being sacrificed in the process. Furthermore, the shift from tundra to shrubs and forests alters the energy budget of the area – for instance, by replacing a lighter surface with a darker surface – making different demands on the hydrology of the region and supporting completely different fauna. Some of these effects actually could enhance or accelerate climate change, while others – such as the shading effect of new shrubs that could prevent permafrost melting and the subsequent release of methane[15] – might partially counteract the effects of the warming trend.

9.1.1.2. Coastal erosion and rising sea levels

Since, as we have seen, rising temperatures lead to the thermal expansion of water and the rapid melting of land-based ice, rising sea levels have become a cause of great concern. Their impacts on cities and human livelihoods aside, rising sea levels have contributed to dramatic shoreline erosion. The result is that previously sensitive habitats, such as the spectacularly productive intertidal zone and tropical mangroves, are being "squeezed" between higher sea levels and uninhabitable inland geography. Beach erosion is also being exacerbated by human activities, such as mangrove clearance, and is being seen from the United Kingdom to Australia.[16]

Inland and coastal wetlands are among the most vulnerable ecosystems on Earth. They serve several critical functions: they are natural barriers to flooding, they purify water, and they are considered some of the most biologically rich of all habitats. They even sink carbon. But altered tidal dynamics, human interventions such as land reclamation for development, and rising sea levels are contributing to the degradation of these ecosystems, as is evident in, for example, Louisiana and the Thames estuary in the United Kingdom.

Consider how human activities are affecting these coastal systems. Dams prevent sediment from reaching deltas and accumulating in rich marshlands. The heavy use of fertilizers changes nutrient cycling in these systems, and causes mangroves to encroach on other wetlands. Wave action, higher sea levels, and storms chip away at coastlines, some of which are left more fragile by bleached coral. Multiple impacts are in play here, not just sea-level rise.

As we struggle to grasp the extent of the climate change impacts that have already occurred, we are also attempting to determine what we can expect to experience in the coming decades. Clearly, this depends on exploring various scenarios of greenhouse gas emissions, the resulting climatic shifts, and a host of other interwoven drivers. The best we can do is to adapt our projections to changing information about our emissions trajectory and our growing understanding of the ways that ecosystems respond to climatic shifts. Even so, there is great certainty that climate change will continue to have a dramatic effect on ecosystems around the world, particularly sensitive and important regions such as tundra, forests, and grasslands.

9.1.2. Impacts in the oceans

Oceans are fundamental elements of the climate system. They modulate the global environment by regulating the climate and through biogeochemical cycling. They are also home to complex ecosystems and myriad species – many of which are of crucial economic and social importance to humans. Forty percent of the world's population lives near an ocean, making oceans the building blocks of our economies and societies. Much like coastal systems, oceans are suffering from several coinciding impacts of climate change. In addition to rising global temperatures and increasing acidity, oceans are influenced by human activities such as overfishing, pollution, and the introduction of non-native species that alter the delicate balance of ecosystems.

As global temperatures rise, so do the temperatures of the oceans. As Figure 9.1 shows, there has been a strong trend in increasing ocean heat content since 1955. The change is not consistent, of course, since seasons, ocean currents, and natural variability all affect the temperatures we find in various places and from year to year.

Corals are some of the first creatures to suffer from this rise in temperature. Coral "bleaching" happens when temperatures rise about 1°C higher than normal for a period of around four weeks. When it gets too hot, corals kick out tiny creatures with which they normally live in symbiosis, turning white and fragile in the process. In the western Pacific Ocean, one of the largest bleaching events occurred in 1998, when 16 percent of the

Figure 9.1. Heat Content of the Upper 700 Meters of the Water Column of the World Ocean, 1955–2008

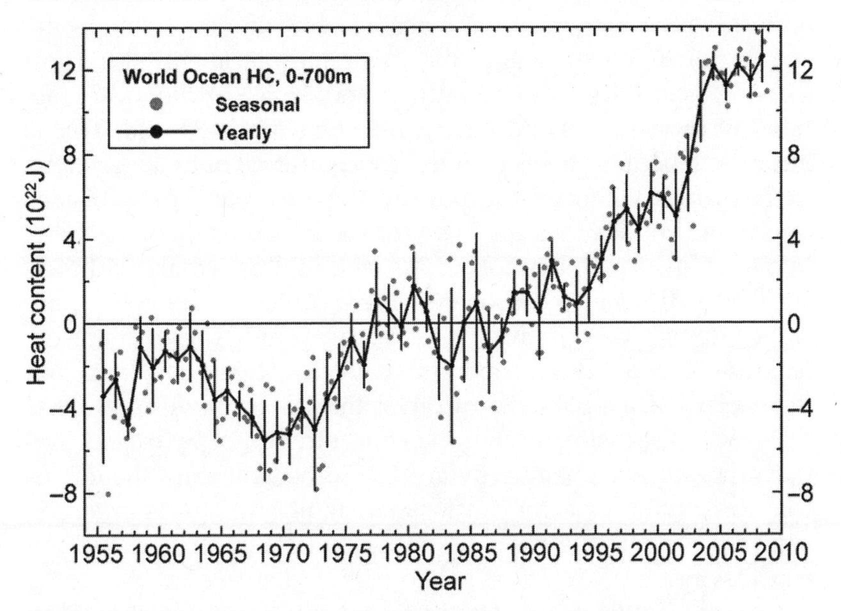

Source: United States, Department of Commerce, National Oceanic and Atmospheric Administration, "Frequently Asked Questions: How Do We Know the Ocean Is Warming?" (Silver Spring, MD: NOAA, 2010).

world's corals died. Corals that remain after a bleaching usually house a different mix of species and are less structurally complex. This is one area, though, where it is difficult to disentangle the effects of climate change from other processes at work in the Pacific Ocean, as its temperatures naturally shift along with the Pacific Decadal Oscillation and the El Niño Southern Oscillation. Overfishing, pollution, and the concentration of CO_2 in the atmosphere also directly affect the health of ocean ecosystems.

When thinking about the effects of the combustion of fossil fuels, what most often comes to mind are the greenhouse gases that are released and the effect they have on the atmosphere. But carbon dioxide affects the planet in another way. Atmospheric concentrations of CO_2 are much higher now than they were before the Industrial Revolution, and rather than just remaining in the atmosphere, much of this CO_2 actually finds a new home in the oceans, where it dissolves in the water and reacts with it to form carbonic acid (H_2CO_3). Through another series of chemical reactions, hydrogen ions and bicarbonate ions are produced. This is important, because the higher the concentration of hydrogen ions in the oceans, the lower (or more acidic) the oceans' pH. This has important implications for many of the creatures that live there. In particular, some plankton, corals, and mollusks that build their shells out of calcium carbonate have a difficult time getting the carbonate they need in these more acidic waters. These organisms form the foundation of food chains in the oceans, and also join together to create habitats for other animals in the form of coral reefs. Scientists estimate that the concentration of hydrogen ions in the oceans has gone up by around 30 percent since the Industrial Revolution, and they anticipate significant effects on ocean ecosystems – although we lack sufficient data so far to draw firm conclusions.

Oceans, however, do not just suck up carbon and moderate temperature fluctuations; they also act like huge conveyor belts, moving heat around the globe. Oceans actually move the same amount of heat around the planet as the atmosphere does. It is this conveyor belt, in the form of the Gulf Stream, that makes Canadians jealous of the relatively mild temperatures in northern Europe despite rough similarities in latitude. This conveyor belt, or thermohaline circulation (see Figure 9.2), is pushed along in part by the differences in density that arise from variations in temperature and salinity in different parts of the ocean. This circulation can be disrupted by, for example, too much fresh water entering the system from melting land ice, which can slow down the conveyor belt. Ocean circulation can also be altered by changes in water temperature. Models show this is actually a very fast process, and the warming associated with a quadrupling of CO_2 in the atmosphere might be enough to cause a rapid shutdown of the Atlantic Ocean circulation that makes it so balmy in Europe. The important lesson here is that ocean circulation changes are nonlinear, and that abrupt

Figure 9.2. The Pattern of Thermohaline Circulation in the World Ocean

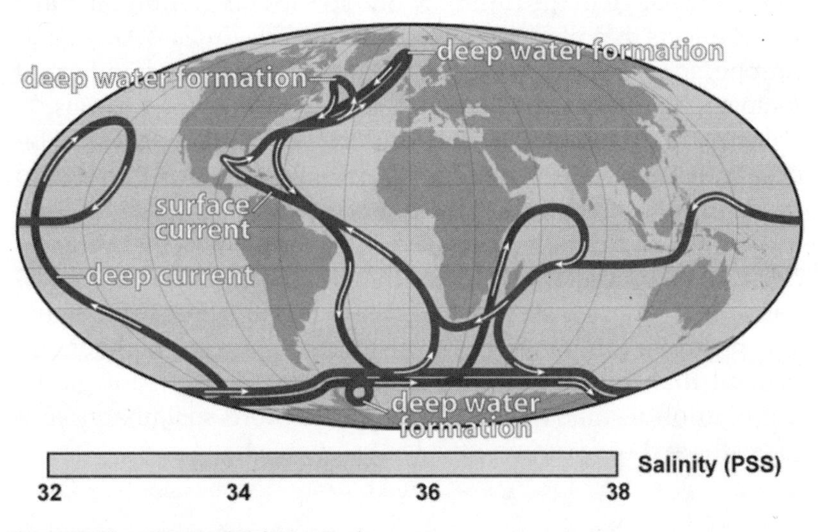

Note: PSS is Practical Salinity Scale.

Source: United States, National Aeronautics and Space Administration, NASA Earth Observatory, Thermohaline Circulation" (Greenbelt, MD, 2008).

changes and thresholds in the system are possible outcomes of a warming climate.

As do species on land, some fish adapt to changing conditions by finding more comfortable places to live. Ocean temperatures do not rise uniformly but in pockets, so scientists are observing changes to the ranges of fish species, many of which form the foundation of our fisheries. As freshwater ice melts and flows into the oceans, their salinity levels decrease. This causes plankton, which are major food sources for larger species, to move out of their traditional territory. Diseases and nonnative species are also popping up in new places. These changes are occurring around the globe.[17]

Changes, moreover, are not just occurring in the oceans. For example, some fish in North American lakes and rivers are migrating anywhere from six days to six weeks earlier than they once did.[18] Why does this matter? Aside from the obvious potential impacts on the fishing industry and livelihoods, each species has its own role to play in a web of interactions with other species, so a change in the timing of one species' migration can have ripple effects that place stress on other species. In East Africa and Europe, scientists have also noticed a decrease in the numbers of diatoms, a form of algae that is the foundation of food chains.[19]

When thinking about the impacts of climate change on oceans, it is important to remember that the natural cycles and processes that affect the oceans are *planetary* in scale. Melting ice or the introduction of invasive species can directly influence ecosystems on the other side of the globe. Although all ecosystems are connected to some degree – through the atmosphere and precipitation patterns, for instance – nowhere is this more evident than in the oceans. This suggests the need for global cooperation to manage ocean resources more sustainably and to evaluate the cascading effects of climate change.

9.2. Adaptation in natural systems

When we speak about responding to the impacts of climate change, we mean adaptation, rather than mitigation. In other words, what capacity do natural systems have to adapt to changing climatic conditions, and what can humans do to enhance this capacity? Approaches to facilitating adaptation of natural systems can range from tiny, local-scale projects, such as pollinating flowers by hand to replace decimated bee populations, to global-scale negotiations about biodiversity conservation. Given the issues of scale we have explored thus far in this chapter, action at both levels might be necessary to slow the massive extinctions for which humans are responsible. We discuss adaptation in human systems in Chapter 10, but here we explore approaches that are not typically thought of as climate change responses, but that nevertheless deeply affect the capacity of ecosystems and species to survive a changing climate.

One of the most important characteristics of ecosystems and species that enhances their ability to adapt to shocks or gradual changes is biodiversity. At the species level, genetic biodiversity – the variety present at the level of genes – enhances the resilience of that species. More genetic variety in a species or a population means a higher likelihood that some individuals will adapt to changing conditions. Lower genetic variety results in uniformity of species, and ultimately translates into vulnerability.

As an example, modern agricultural practices typically are monocultures – the practice of planting vast swathes of genetically identical plants. This is an advantage when it comes to growing and harvesting crops, but it can be a problem when a disease or parasite attacks the field, as every plant in the field will be susceptible. Monocultures are also unable to deal well with changing conditions, such as the changing precipitation and temperature regimes associated with climate change.

Curious characteristics emerge, however, in genetically diverse communities, allowing some species to thrive even if conditions rapidly change. In Costa Rica, for instance, a particularly unusual vine, *Marcgravia evenia*, has evolved with peculiar concave leaves that are just perfect to capture and return the strong echolocation signal that bats send out.[20] Imagine varieties of this vine that also exist with more traditional flat leaves. Now imagine that a particular bird and a bat are both attracted to this vine and provide valuable pollination services, but a disease or extreme event wipes out the bird population, leaving only bats. Suddenly, the clever satellite-dish variety becomes more viable than its flat-leaved counterpart, and it is this diversity in form that allows the vine ultimately to survive the shift in the abundance of pollinators.

Species biodiversity, in contrast to genetic biodiversity, refers to the number of "groups" of organisms, each group consisting of individuals similar enough to interbreed or exchange genes. This is much easier to measure than genetic diversity, because it can be done without a lab. Each species has a role in an ecosystem, so the loss of a species can affect the ecosystem as a whole. For this reason, the bulk of biodiversity conservation work has focused on preserving species.

Ecosystems and species do their best to adapt naturally to changing conditions, but humans are also taking steps to undo some of the harm that has been done. Biodiversity conservation originally focused squarely on preserving species that humans enjoyed. In the United States, for instance, the system of national parks was created by a group of wealthy and influential fishers and hunters. Conservation for human use has now broadened, though, to include the preservation of species that are not directly related to our own enjoyment and profit, and to acknowledge that ecosystems are complex systems and that the species within them play a multitude of carefully balanced roles. Areas (rather than species) are protected if they are considered biodiversity hotspots or if they hold particular cultural or social value. Protected areas range from the lowland forests of Madagascar to the Rift Valley lakes in East Africa to parks along the Colorado River in the United States.

Choosing which species to preserve is a complicated task that involves evaluating how climate change and other pressures are affecting species, and identifying species that are most vulnerable and those that are most valuable (both to humans and to the functioning of the ecosystems of which the species are a part). One way that we do this is by evaluating *ecosystem services*.

Humans benefit from the healthy functioning of ecosystems in a variety of ways. We breathe in the oxygen created by photosynthesizing plants. We grow crops that consume soil nutrients created through decomposition and nutrient cycling. The natural world also offers opportunities for rest, reflection, and recreation. Many ecosystem services are "public goods": your use of the service (say, enjoying a beautiful view of a temperate rainforest) does not affect my use of it (I can still enjoy the view as well, since it has not been "used up"). The "cultural services" that ecosystems provide often fall into this category. Other ecosystem services, however – such as provisioning services in which ecosystems give us products to consume – are not so win-win. A tract of forest provides only so much wood over a given period. If I use the wood, you cannot use it also. In these cases, individuals and groups need to work together to manage the resource cooperatively, or to assign property rights, to ensure that the resource is not overconsumed or damaged.

Ecosystem services are often affected by externalities – the unintended (and uncompensated) side effects of our activities – as is the climate more broadly. In many countries, companies do not pay – for instance, through a carbon tax or a cap-and-trade system[21] – to emit greenhouse gases. For them, dumping carbon dioxide into the atmosphere is totally free; in other words, the value of the services that the atmosphere provides is not sufficiently captured, the result of which is a warming climate and threatened ecosystems. The capacity of ecosystems to provide valuable services, in turn, is gradually (and sometimes rapidly) being diminished. In a now-classic effort to demonstrate that ecosystems have a real and measurable value, a group of scholars has put a price on the value of sixteen different ecosystem services in sixteen biomes, or large naturally occurring areas that contain a particular mix of flora and fauna.[22] They suggest that, although annual global gross national product at the time of the study (1997) was approximately US$19 trillion, the annual value of ecosystem services was US$33 trillion. Critics argue, however, that there are significant gaps in the data were used to construct this estimate, and that the ecosystem value is so high as to render it nearly meaningless in the context of willingness (or even ability) to pay. Nonetheless, it is certainly the case that most of the value of ecosystem services remains outside the market, and is vulnerable to inequitable consumption and severe externalities.

Some of these problems seem overwhelming, but interesting actions are being taken around the world to preserve species, protect coastlines, and address the impacts of climate change. Collective action to preserve biodiversity takes many forms, most of which have nothing to do with assigning private property rights. Biodiversity reserves, or protected areas, for instance, are defined spaces within which particular activities are banned. An example is the Galapagos Marine Reserve, which covers 133,000 km^2 of ocean and islands off the coast of Ecuador. This reserve is home to more than 2,900 different species, including the Galapagos tortoise, whale sharks, iguanas, dolphins, and sea turtles, and encompasses diverse habitats ranging from coral reefs and mangrove swamps to underwater cliffs and lagoons. The purpose of the reserve is to ensure sustainable use of the area's resources while encouraging education, research, and responsible tourism.

Threatened and endangered species designations are another tool that can ensure the conservation of biodiversity and, ultimately, enhance the resilience of ecosystems to climate change. Threatened species are those likely to become endangered if proper protections are not put in place, while endangered species are those that must be protected if they are to survive, or else they will become extinct. The International Union for the Conservation of Nature, the main global authority on the conservation of species, compiles a Red List that is the most comprehensive assessment of the status of biological species on the planet;[23] a multitude of country-specific lists exists as well. The main purpose of labeling a species as at-risk is to trigger national programs for species protection or global treaties focused on conservation. Threatened or endangered species designations have been the most common method of protecting biodiversity, but more integrated ecosystem-wide approaches are also emerging.

Other steps also can be taken to tackle the multiple impacts of climate change on coastal systems. Managing the use of fertilizers and controlling runoff help to prevent elevated levels of nutrients in these systems. Preserving the intricate mangrove forests that are found on nearly three-quarters of tropical coastlines can help buffer coasts from waves and storms. By creating protected areas and making sure that animals are able to safely migrate, breed, and feed, humans can help to ease the stress of a warmer climate. In the next section, we explore in more detail policies and actions to address these multiple and conflicting stressors.

9.3. Policy tools and progress

Policy efforts are under way to conserve biodiversity and to help ecosystems become more resilient in the face of a changing climate. Many of these efforts are international in scope. Indeed, a cross-border approach is vital: just four countries, for instance, contain two-thirds of all of Earth's primate species, and given the globalized economy and the need to preserve resources for future generations, it is in the interests of people outside these countries to help such efforts. The sections that follow present a sample of policies and ongoing negotiations that facilitate a coordinated approach to biodiversity conservation.

9.3.1. International tools

Climate change is a problem that affects multiple areas of human development and environmental integrity, so that not all of its impacts can, or even should, be dealt with through policies specifically targeting climate change. Many scholars would argue that the only effective way to address climate change is to embed these concerns throughout a wide variety of policy areas. One such area is emerging approaches to biodiversity conservation and management.

Until the 1980s, the focus was on individual species, and almost never on habitat preservation and never coordinated. Our understanding of the scale and irreversibility of biodiversity loss was growing, however, and a global regime seemed the best solution since, although species are not evenly distributed around the planet, biodiversity came to be regarded as a shared global resource. In 1988, ad hoc working groups were established to lay down the foundations of a Convention on Biological Diversity (CBD), which was presented for signature at the 1992 United Nations Conference on Environment and Development in Rio de Janeiro, Brazil (the Rio Earth Summit). The Convention, which entered into force in 1993, has three main goals: the conservation of biodiversity, the sustainable use of biodiversity, and an equitable sharing of benefits. The main issues under the Convention are

- strategies and incentives for conserving biodiversity;
- regulated access to genetic resources and traditional knowledge – that is, learning from those who have used the resources for centuries, and obtaining prior informed consent from the party that is providing the resources;
- access to and transfer of technology to governments and/or local communities that provide the traditional knowledge and/or resources;
- technical and scientific cooperation; and
- the raising of public awareness.

As with any international treaty or convention, however, *domestic* tools are required to implement international regimes. In the case of the CBD, these tools are National Biodiversity Strategies and Action Plans (NBSAPs), which signatory countries are required to create; to date, 91 percent of the parties to the CBD have done so. There are also plans to integrate biodiversity into domestic poverty reduction and other sectoral and cross-sectoral policies. Yet, despite the prevalence of NBSAPs, a recent assessment of national-level efforts paints a bleak picture of the progress made so far.[24] Few countries have established time-bound, measurable targets – a critical element of translating high-level plans and promises into actions that can be evaluated and altered. Although actions to preserve biodiversity and adapt to climate change have been designed, countries rarely prioritize among the actions, and few have developed effective mechanisms for monitoring and review. Largely absent are strategies for communication and financing, which are required to take planned actions and accelerate the uptake of strategies in other regions. Subnational strategies and action plans are another necessary next step, as plans at the national level must be implemented on the ground in regions and communities. This is the challenge of planning for biodiversity in a changing climate: since ecosystems cross borders and boundaries with abandon, a delicate mix of local implementation and global or multilateral collaboration is required to produce the most robust and desirable outcomes.

9.3.2. National and subnational tools

Communities and nations around the world are struggling with the task of achieving multiple objectives simultaneously. Should greenhouse gas reductions occur at the cost of biodiversity? Should we adapt to climate change by creating neighborhoods that are more like bunkers than thriving, beautiful, livable spaces? In addition to the national-level plans that have been developed in association with the CBD, one emerging set of solutions that holds significant promise is ecosystem-based approaches, which could deliver on multiple priorities simultaneously, rather than just one. Ecosystem-based approaches mean using living systems, such as forests and wetlands, to do the work of traditional "gray infrastructure." An example is using a series of salt marshes or wetlands to capture and purify storm water, rather than burdening sewage systems and risking floods. Another example is using oyster reefs to protect coastlines from wave action or storms, rather than simply building higher seawalls. In the sections below, we explore a couple of examples of ecosystem-based approaches in practice.[25]

9.3.2.1. Ecosystem-based approaches at work: The Wallasea Island Wild Coast project

The Wallasea Island Wild Coast project is a scheme proposed by the United Kingdom's Royal Society for the Protection of Birds in a largely rural area northeast of London, along the Crouch and Roach estuaries in the county of Essex, and is a coastal habitat restoration and climate change adaptation project on a scale unique in the United Kingdom. The geographic and ecological context of the project is already one of change: the land originally was reclaimed from the North Sea more than four hundred years ago and converted to agricultural land. Despite the construction of substantial flood defenses – typically considered "gray" infrastructure – in the area, the UK Environment Agency recently determined that they were no longer economically viable, meaning that continued public expenditures on them was unlikely. This decision puts the surrounding 12,100 ha – an area the size of about 23,000 American football fields – of floodplain at risk of unmanaged inundation, especially given predicted sea-level rise of 6 mm per year in this region.[26]

The Wallasea Island Wild Coast project consists of four major components:

- the purchase of 744 ha of mainly low-productivity agricultural land on Wallasea Island;
- land-raising and landscape engineering by importing material from major infrastructure projects elsewhere in England (such as the Crossrail and Thames tunnel project);

- managed realignment – controlled breaches of the seawall around Wallasea to allow inundation by the sea and the creation of new biodiversity-rich habitat; and
- development of public education and recreation opportunities to provide socio-economic co-benefits.

The two central aims of the project are to offset historical losses of coastal habitats on the island and across the rest of Essex and to address the ongoing and growing flood risks in the region. Adaptation to climate change impacts, including rising sea levels and altered habitat suitability, was a major consideration during the planning of the project. In addition, the potential to transform the land from a net source of carbon emissions to a net sink was explored. The newly restored landscape will be a wetland mosaic of mudflats, saltmarsh, lagoons, and pastures. These will support nationally and internationally important bird populations, with the hope of re-establishing lost breeding populations of birds such as spoonbills and Kentish plovers.

The green infrastructure approach used in this case carries with it several advantages compared to other options. A more traditional gray-infrastructure approach might employ regulated tidal exchange, while no response at all would permit an unmanaged breach of the flood protection system. Instead, a green infrastructure approach fosters the strategic creation of crucial habitats, and offers public education and recreation benefits. Furthermore, this project sets an exciting precedent: waste material from other infrastructure projects can be used to reach both adaptation and mitigation goals.

9.3.2.2. Ecosystem-based approaches at work: Peatland rewetting in Belarus

A very different suite of ecosystem-based strategies is currently being implemented in the eastern European nation of Belarus, where peatland covers approximately 2.4 million ha – an area more than half the size of Switzerland. Around half of this area has been affected by drainage and peat-extraction activities, so that the fragmentation suffered by these damp habitats and their local species has been extensive. The project addresses the potential of carbon emission reductions and ecosystem restoration from rewetting degraded or depleted peatland. In increasingly dry conditions, peatland that is typically saturated with water becomes exposed to air. As a result, organic material begins decomposing rapidly, releasing carbon dioxide into the air. In healthy peatland, this process is slowed by the anoxic (oxygen-free) conditions produced through saturation. The aim of the project is simultaneously to restore and sustainably manage a large area of peatland, while sinking extraordinary amounts of carbon to create quantifiable emissions reductions – in other words, the project aims at both climate change adaptation *and* mitigation.

There are a couple of key ways this project ensures that the mitigation benefits are reaped, alongside biodiversity conservation goals. Project developers are verifying and quantifying emissions reductions, packaging them into units that can be traded (or purchased) on the voluntary carbon market. The restoration of these habitats is also helping to re-establish ecosystem functions and form ecological corridors and reservoirs, allowing for the migration of species and the enhancement of their populations.

Currently, the project has rewetted six depleted or degraded peatland sites covering more than 9,000 ha of restored land. The ecosystems in these areas are now re-establishing their functions, and some are starting to yield ecosystem services such as food, microclimate regulation, and landscape enhancement. Water levels and the vegetation in the rewetted sites are being closely monitored both to maintain the ongoing restoration process and to assess the levels of greenhouse gases being emitted. This project represents one set of strategies for bringing social, cultural, and economic benefits to local communities, and to Belarus as a whole by enhancing the capability of its ecosystems to provide goods and services in a sustainable manner.

9.3.2.3. Ecosystem-based approaches: Conclusion

These two cases (and there are many more) show that ecosystems-based approaches to climate change adaptation and mitigation hold significant potential to deliver on multiple priorities simultaneously. The European Union, along with parties to the Convention on Biological Diversity, has stressed the importance of these approaches, providing tools and resources to support their implementation.[27] This further highlights the need to cast our net wider than simply over climate change – to come up with solutions that strengthen ecosystem integrity, enhance community livability, and trigger a shift towards more sustainable development pathways.

9.4. Conclusions

The picture of the key impacts of climate change on Earth's natural systems is a complicated one. Rising temperatures and higher concentrations of CO_2 in the atmosphere are affecting corals, fish species, and even ocean circulation patterns. Forests are migrating in response to changing temperatures and precipitation patterns, adding stress to threatened species and degrading ecosystem services. Some scientists think that this is bringing humans to a tipping point, after which the oceans and other ecosystems might function very differently. Although the issues we have discussed in this chapter do not represent the full range of effects the planet is experiencing and is expected to experience, they do allow us to gain some idea of the breadth of challenges that natural systems face. Perhaps most important, a range of human activities (not just the emission of greenhouse gases) is exacerbating the impacts of a changing climate, yielding unintended consequences. Despite this rather bleak picture, ecosystem-based adaptation and mitigation, along with the incredible capacity of some ecosystems to adapt autonomously to climatic shifts, hold out the possibility that remedies are at hand.

CHAPTER TEN

Climate Change Impacts on Human Systems

10.1. Introduction
10.2. Key concepts in climate change impacts and adaptation
10.3. Observed and projected impacts of climate change

- 10.3.1. Impacts on water and food
- 10.3.2. Impacts on cities and infrastructure
- 10.3.3. Equity implications: Health and the global distribution of wealth

10.4. Adaptation in human systems

- 10.4.1. How to "do" adaptation

10.5. Policy tools and progress

- 10.5.1. Policy tools for adaptation
- 10.5.2. International and national adaptation
- 10.5.3. Subnational adaptation
- 10.5.4. Social movements and human behavior: The root of the adaptation conundrum

MAIN POINTS:

- Climate change impacts refer to those phenomena that occur as a result of rising temperatures and shifting precipitation patterns. These impacts can be either positive or negative.
- Observations from around the world indicate that impacts are already occurring, ranging from increasing drought to rising sea levels and shifting patterns of ecosystem distribution.
- These natural impacts have human implications, such as reduced availability of drinking water, food shortages, floods in low-lying communities, and prolonged heat waves.
- Our vulnerability to climate change impacts is determined by a variety of factors, including our exposure to particular impacts and our capacity to adapt to them.
- Adaptation can be either proactive or reactive. We have a long history of reactive adaptation, but proactive adaptation requires a better understanding of the scale and distribution of climate change impacts as well as new and effective response strategies.

10.1. Introduction

In 2012, two major sawmill explosions shook the towns of Burns Lake and Prince George, in the Canadian province of British Columbia. Four people were killed, two in each incident, and many more were injured. Was climate change to blame?[1] This might seem like an absurd conclusion – a classic case of using climate change to explain any and every bad thing happening in world at any moment. In this case, however, it just might be true. Consider this: climate change is causing warmer winters in British Columbia. Because of these warmer winters, paired with forest-management practices that suppress fire and leave older trees standing, a common pest, the mountain pine beetle, is thriving, killing vast swathes of pine trees across the province. The mills in question were processing the dry, resin-filled wood killed by pine beetles, rather than the green wood of healthy trees that typically arrives. This causes columns of extremely flammable sawdust particles to float through the air of the mill, creating a tinderbox that will ignite at the slightest spark.

Following the chain of causation from greenhouse gas emissions to impacts on humans can be a tricky business. Impacts such as damage to buildings along a coastline often result from a constellation of factors. In the case of hurricane damage and loss of life, for instance, the increasing severity of hurricanes is only part of the problem. The growing number of homes and businesses built on vulnerable coasts, the effectiveness of emergency management plans, and the awareness of people who live and work in these areas all have a bearing on the ultimate human toll.

For many, the most critical impacts of climate change are those that directly affect us, not those that affect natural systems and species distributions. There are many reasons for this focus. We are only now beginning to estimate the financial value of the services that ecosystems provide, and even if we are successful, a dollar figure often fails to capture the cultural or spiritual significance of ecosystems. Many of us are also physically and emotionally distant from the natural world – we live in cities, buy our food in grocery stores, and vacation at resorts.

It is thus the human dimension of climate change that has stimulated action in many quarters. In this chapter, we look at the range of impacts of climate change in both developing and developed countries, and explore the ways climate change is already affecting food systems, the availability and quality of water, infrastructure and housing, and a host of other issues. We also examine the influence that poverty and global inequality has on our vulnerability to climate change. Perhaps most crucially, we uncover key strategies that can be used to respond to the challenges climate change presents.

10.2. Key concepts in climate change impacts and adaptation

Before we delve into the climate change impacts that are already being observed, and those that are expected, in human systems, it is important first to develop precise definitions of three important, and interrelated, concepts: vulnerability, adaptive capacity, and resilience. Though distinct, these concepts overlap and can contain feedbacks. For example, a community's capacity to adapt influences its vulnerability, which, in turn, feeds back to further influence its capacity to adapt.

We are only now beginning to grasp the extent of the impacts of climate change, and significant uncertainty remains. The jagged line in Plate 4, for instance, shows the observed temperature record from 1900 to approximately 2005. Stretching out from this line are two smooth lines that encompass the universe of scenarios that scientists have generated to help us understand what level of warming we could expect in a high-emissions future versus a low-emissions one. Many scientists believe that humans are already committed to at least 2.0–2.5°C of warming, regardless of how quickly we reduce emissions of greenhouse gases. The worst-case scenario leads to accelerated warming beyond our control and beyond the capacity of natural systems to adapt. At each level of warming, scientists expect to observe significant impacts from climate change, including potable water shortages, declining crop yields, and flooding in low-lying areas.

In Chapter 9, we looked at various dimensions of vulnerability in the context of natural systems, including the magnitude, timing, and probability of climate change impacts. But what about the human determinants of vulnerability? Plate 5 shows the distribution of vulnerability[2] around the globe in a high-temperature future when both sensitivity and exposure to climate change impacts are taken into account. China, India, and large parts of Africa and the Middle East are likely to be exposed to significant impacts in this scenario and are also very sensitive to them. This is especially true when we consider the capacity of some of these developing nations and emerging economies to adapt to climate change.

Why exactly are developing nations more vulnerable? Three leading adaptation scholars, Nigel Brooks, Neil Adger, and Mick Kelly have found that vulnerability to climate change is deeply influenced by such factors as maternal mortality, access to sanitation, calorific intake, civil liberties, government effectiveness, and the female-to-male literacy ratio. These are clearly linked to one another. High levels of literacy, for instance, allow individuals to lobby for civil liberties, which, in turn, enhance government effectiveness.[3] If a storm causes a breach in sewage facilities and contaminates water, fewer people will die if the population is healthy and well fed. In other words, vulnerability to climate change is directly related to the underlying patterns of development. We take these issues up again at the end of this chapter, when we touch on issues of adaptation.

The simplified diagram in Figure 10.1 illustrates that true vulnerability is also influenced by adaptive capacity, an important concept when it comes to the human dimensions of climate change. Of course, the relationship between these concepts is not as clear as depicted here. Since all of these elements feed into determining our ultimate vulnerability, there is significant overlap among them.

Figure 10.1. Determinants of Vulnerability to Climate Change

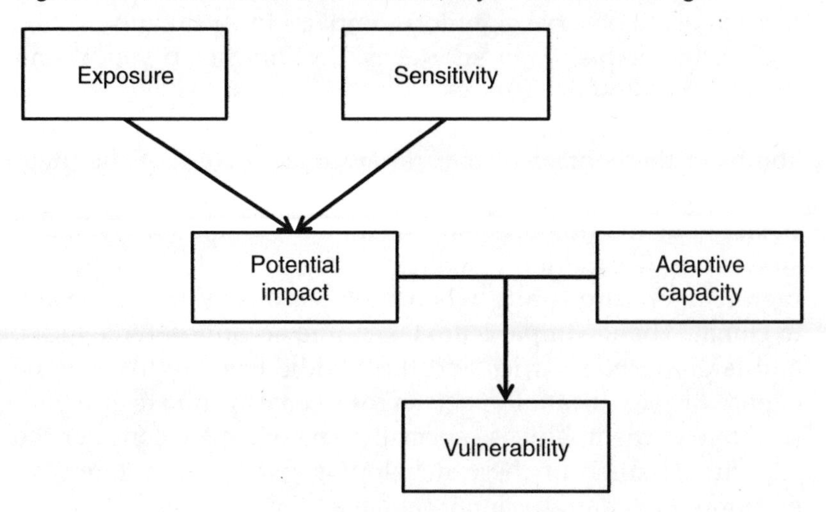

Adaptive capacity is an important element of vulnerability, and is usually used to refer to human systems, rather than natural ones. The ingredients of adaptive capacity appear to be

- the availability and distribution of resources such as food and energy;
- technological options for adaptation;
- human capital (such as education);
- social capital (networks of trust and collaboration among groups);
- mechanisms, such as insurance, that spread risk;
- the ability of decision-makers to obtain, manage, and respond to information; and
- the public's perception of the cause of climate change effects, and responsibility for acting.[4]

Some of these factors clearly overlap with the determinants of vulnerability we discussed above, but they pertain more specifically to the challenge of adaptation – or responses to the impacts of climate change. In other words, some of the same factors that determine how vulnerable we are to climate change also shape our capacity to respond to this challenge.

Adaptive capacity, however, still represents a *potential* to respond to climate change. A multitude of factors intervene and influence actual decisions to use scarce resources to adapt to climate change. Thus, adaptive capacity influences vulnerability in a very real way, but does not necessarily tell us whether a region or system will, in the end, suffer the impacts of climate change.

Another term that is often heard in discussions of climate change impacts and adaptation is *resilience*. Adaptive capacity generally refers to the ability of human systems to respond to climate change, but resilience is often used to refer to the ability of a system to return to a healthy state following a change or shock, even if this new state differs from the state the system was in before the disturbance. For instance, if a community relies on water from a glacier-fed river that is gradually running dry as the climate warms, it might adapt to this change by capturing rainwater to fill the same need. Such an action, however, requires an understanding of changing precipitation patterns, altering the water distribution system to accommodate rainwater capture, and other elements of adaptive capacity. If these elements are present, the community might be considered more resilient to climate change if it is able to return to the "healthy state" of having enough water for drinking and sanitation, even though it has adapted to obtain it from a different source. The resilience of a system thus depends on its coping range, or the amount of change it can absorb without fundamentally shifting. Adaptation helps to expand this coping range, raising the threshold beyond which the system changes, and enhancing resilience, making the system less vulnerable to the impacts of climate change.

Uncertainty remains high, however, in our assessment of impacts, for several reasons. First, many impacts are contingent on what we decide to do about greenhouse gas emissions. Clearly, this is still in negotiation, and we are not likely have a clear picture of our emissions future any time soon. Second, scientists are beginning to explore the possibility that we will reach major thresholds beyond which warming accelerates uncontrollably and the planet is fundamentally altered. We do not yet understand enough about these thresholds, or if we are reaching one, and what our options are if we do. Third, and most obviously perhaps, natural systems are highly complex and interwoven. Scientists and other experts do not understand completely the nature of ecosystems' coping ranges, and what can be done to enhance resilience. As such, multiple scenarios of impacts must be produced to help communities understand the various futures they might be facing.

Uncertainty grows as we move from emissions scenarios to the response of the global climate and its impacts on systems. Many scientists argue, however, that we need to get comfortable with uncertainty, since we will never be able to predict flawlessly the behavior of human and natural systems. Nevertheless, climate change mitigation and adaptation are essential.

10.3. Observed and projected impacts of climate change

10.3.1. Impacts on water and food

Closely connected to the socio-economic and human health issues that we describe later in the chapter is the issue of the impacts of climate change on the availability of natural resources. One of the most significant impacts, for instance, is on the availability of drinking-water supplies. In 2010, the United Nations identified access to safe fresh water and sanitation as essential to the realization of all other human rights.[5] The right to water is also embedded in the Millennium Development Goals, the UN's blueprint for enhancing human well-being.[6]

Access to water is not the only problem. Extreme precipitation, leading to flooding, might increase the pollution of drinking-water reservoirs, putting populations at risk. This is especially crucial in countries with underdeveloped sanitation and sewage systems that are prone to overflow or breach. In other words, the vulnerable can be made more vulnerable. Although precipitation will increase in some areas, increased evaporation and more intense and prolonged droughts will create added stress in others. Often, these effects are quite regional. For instance, if a large wetland is unsustainably managed and drained, less water will be available for evaporation in the region. This mighty directly affect regional-scale precipitation patterns, exacerbating drought conditions in an amplifying feedback of the type we discussed in Chapter 2.

The core climatic trends that are affecting freshwater supplies are the following:

- Earlier spring and warmer summers, leading to shifts in the peak flow of rivers that are affected by snow or glacier melt. Often, this takes the form of an earlier, more dramatic "spike" in stream flow, creating challenges for the sustainable management of water resources.
- Higher stream flows at first, as glaciers melt more quickly, followed by continually diminishing stream flow as the glaciers disappear permanently.
- Degraded water quality as warmer temperatures drive algal blooms and "thermal pollution" – or warmer water in ecosystems that traditionally have cooler water – leading to the deaths of fish and other species that are important for humans.[7]

We do not have to wait to see what climate change will do to our abundant supply of clean drinking water – it is happening now. Clear evidence shows that precipitation is increasing in latitudes north of 30° and decreasing in latitudes from around 10°S to 30°N, leading to more intense droughts.[8] Glaciers, which feed the world's most heavily exploited rivers, are decreasing almost everywhere, and permafrost is thawing.[9] These observed impacts make a compelling case for mitigation of climate change – getting at the roots of the problem – but they also suggest that proactive, effective adaptation is required. Of course it is not just climate change that is affecting the availability of safe freshwater supplies: also crucial is the demand for particular foods. The rise of industrial agriculture, paired with an escalating demand for the typically meat-heavy Western diet, puts incredible strain on the planet's freshwater resources while loading rivers with pesticides and fertilizers.

Climate change is also affecting the abundance and distribution of plants and animals that provide humans with food, livelihoods, and fuel. Interestingly, climate change might be a temporary boon for the high latitudes in terms of food production. Longer growing seasons and the fertilizing effect of increasing levels of carbon dioxide in the atmosphere for some plants might lead to higher yields in countries such as Canada. This is the case, however, for only a *small-to-moderate* increase in temperature – that is, between 1°C and 3°C;[10] all bets are off if we go past that. In the middle and equatorial latitudes, even a further small increase in global temperatures might be catastrophic for food production.

Food production is influenced by variables such as the demand for various products, the availability of water, the cost of inputs such as seed, labor, and fertilizer, precipitation events, soil quality and nutrient levels, and a host of others. For this reason, climate change might yield surprises that we did not predict. Take increased drought in some regions, a longer growing season in others, increasing severity and frequency of precipitation events, and an increasing incidence of heat stress. Combine these with increasing pest outbreaks and changing fire risks, and the general consensus is that our global food production system is in for a shock.[11]

If we narrow our focus a bit, however, it becomes clear that small farmers in rural, developing countries are the most vulnerable to the impacts of climate change. Many individuals and communities in developing countries have limited capacity to adapt. They typically have few financial resources to invest in new technologies to enhance productivity under changing conditions or to help to ride out a particularly bad year or steadily declining conditions, and limited access to the clean water that is an integral ingredient in any successful agricultural operation. Furthermore, they often lack other ways to earn an income. Low literacy rates and challenging political conditions create barriers to obtaining work outside traditional small-scale agriculture. Thus, the impacts of climate change on humans and possible adaptation strategies are as much a function of the underlying level of development as they are a function of climatic factors such as precipitation and temperatures.

10.3.2. Impacts on cities and infrastructure

The second way in which climate change directly influences well-being is through the increasing severity of natural disasters such as hurricanes, extreme river floods, and super-cell thunderstorms. The costs incurred and damages that result from these events depend not only on their severity, but also on the value of property and the number of human lives in their path. Human communities have tended to develop in extremely vulnerable areas, such as coastlines, and now are suffering from these exacerbated effects. These findings, however, are contentious. Extreme events are relatively rare and data are sparse. The Intergovernmental Panel on Climate Change, for instance, has "low confidence" that there has been a longer-term – that is, forty-year – increase in tropical cyclone activity, and only medium confidence that some regions are experiencing longer and more intense droughts, though it nevertheless declares it likely that human emissions of greenhouse gases are driving these effects.[12]

It is certainly the case, however, that the concentration of people in cities has a significant effect on vulnerability to climate change. Extreme events are clearly more likely to cause loss of life in cities than in sparsely populated rural areas. As we saw with Hurricane Katrina's devastation of New Orleans, even cities in wealthy countries are vulnerable to weather and climatic extremes. The billions of dollars invested in city infrastructure dramatically increase the cost of extreme events, and city-dwellers are also often disconnected from the source of food, making such communities more vulnerable to disruptions in the food system. Finally, centralized urban systems of energy, freshwater provision, and sewage disposal are extremely susceptible to large-scale malfunctions than more resilient, decentralized systems.

10.3.3. Equity implications: Health and the global distribution of wealth

The third way in which climate change affects human populations is deeply linked with the first two. The economic and social costs of extreme events and their effects on human health might worsen underlying socio-economic issues such as poverty, inequity, and illiteracy. These impacts, in turn, make populations that are less well equipped with the financial and human resources to protect themselves more vulnerable to climate change. Poorer populations are also often more directly dependent on climate-sensitive resources such as local food and water supplies, magnifying the effect of local weather extremes and changing climatic patterns over time. Thus, human populations in the developing world have a considerably lower adaptive capacity than do their counterparts in developed countries. As well, many of the world's poorest people are located in regions that are extremely physically vulnerable, such as the drought-prone regions in sub-Saharan Africa, the flood plains of Bangladesh, and Caribbean islands in the path of hurricanes. It is also these most vulnerable populations that are the least responsible for the current state of climate change.

Although in some cases the impacts of climate change might be similar in both developing and developed countries, the latter are far better able to respond. But even the wealthy industrialized world is unlikely to remain insulated against the worst of the human impacts of climate change for long. Some scientists estimate that climate change could create more than 300 million environmental refugees by the middle of this century. These are individuals whose livelihoods will have been lost to drought, extreme weather events, food shortages, and the conflict brought on by these impacts. Waves of these new forms of refugee then might seek to live safely in countries that have not borne the brunt of climate change. Thus, developed countries should expect to be forced to address tough questions regarding the ethics of denying livelihoods to those who have lost theirs as a result of greenhouse gas emissions from those same developed countries in which refugees seek asylum.

Changing climate also affects human health in several ways. First, rising summer temperatures increase the likelihood of heat-related mortality. In some countries, this is offset by an increase in the use of air conditioning, although this might produce higher greenhouse gas emissions arising from greater demand for power. Second, rising temperatures can affect the spread of diseases, including the geographical range and abundance of vector-borne diseases such as tick-borne encephalitis and Lyme disease, some food- and water-borne diseases, and pollen and dust-related diseases. It is also important, however, to consider other drivers of these altered disease patterns, such as changing land-use patterns – human encroachment on land that supports an abundance of ticks, for instance – and changing human behavior. Although rarely discussed, researchers are also now beginning to consider the effects of climate change on mental health. The potential for trauma, anxiety and impacts on community well-being might become significant in the future,[13] especially in the context of conflict over resources or energy. Responding to climate change will involve a careful consideration of these multiple pathways of impacts.

At the same time, little conclusive evidence exists for the impact of climate change on human health, due to the lack of long-term epidemiological studies and to the powerful influence of non-climatic drivers of disease. The reality is that an intricate web of factors influences the distribution and abundance of disease in human populations, and climate is only one trigger among many. As such, we might never be able to attribute a particular portion of deaths or incidences of illness specifically to climate change. Nevertheless, evidence is building that climate change is exacerbating other problems, such as poverty, which, in turn, affect human health.

Urban environments have always presented unique challenges to human health, and it is expected that climate change will only increase these challenges. During the heat wave of 2003 in western Europe, for instance, it is estimated that more than 52,000 people died as a result of heat stress.[14] Many of these were vulnerable populations, such as the elderly, who were unable to obtain reprieve from the exceptional heat in their city apartments. Due to the urban heat island effect – the result of concentrated dark paved surfaces and waste heat from energy usage – cities are a couple of degrees warmer than surrounding areas. This tends to increase air-quality problems and related respiratory illnesses. As temperatures rise, we also expect to see a shift in the incidence of food- and water-borne diseases, with disproportionately high mortality in urban areas, particularly in developing countries. Similarly, infectious diseases tend to spread more easily in cities, where contact with carriers can be frequent.

10.4. Adaptation in human systems

Adaptation to climate change is generally thought of as a way of managing the risks associated with climate change. As such, the experts in this field often specialize in disaster management, risk perception, geography (and related fields such as urban planning), and ecology.

It is commonly said that adaptation is essentially local in focus, while mitigation is global, but this distinction, in fact, might be false, or at least only partially true. Although it is true that, generally speaking, erecting, say, a flood-protection system protects only those people in the nearby floodplain, we have also seen how the impacts of climate change can reverberate around the globe. Carrying this example forward, if the dike is not constructed and vulnerable populations lose their homes and livelihoods, they might become environmental refugees, which, in turn, could have a very real impact on safe or nearby nations.

Mitigation is also not simply global, as it might appear on the surface. Although greenhouse gas emissions in any location mix in the atmosphere and affect the global climate equally, mitigative responses must be designed, implemented, and paid for locally. So mitigation has an effect on local communities and might also yield co-benefits such as improved air quality.

10.4.1. How to "do" adaptation

Adaptation experts delineate between "reactive" and "proactive" adaptation, in part to identify the ways in which we must advance our knowledge and practice to address an unpredictable future. Adaptation can be viewed as reactive in either stimulus or form. By stimulus, we mean the driving force behind the adaptive action. If scientists or farmers observe that water is becoming scarcer, they can develop irrigation systems that address this problem. Adaptation that is reactive in form refers to actions that do not actually prevent the effect from occurring but help affected groups to recover afterward. Insurance is a good example of adaptation that is reactive in form, as it helps to compensate victims of climate-change-related incidents, but does not protect them physically.

Generally speaking, reactive adaptation is informed by direct experience, rather than by predictions or forecasts. The benefit of this approach is that it is often quite clear how resources should be used to protect against the impact, and uncertainty is low. Reactive adaptation is familiar to human communities – we have been doing it for thousands of years. Good examples of reactive adaptation to climate among human civilizations abound. Crop diversification, irrigation, and water management, for instance, have typified agrarian societies since the development of agriculture nearly 10,000 years ago. By 5000 BC, for instance, the Sumerians (an ancient civilization residing in present-day Iraq) had developed large-scale intensive cultivation of land and organized irrigation. Archaeological investigation has identified evidence of irrigation in Mesopotamia, Egypt, and Iran as far back as the sixth millennium BC, with barley grown in areas where the natural rainfall was insufficient to support such a crop. More recently, humans have developed new reactive adaptation practices, such as disaster-risk management, that help communities recover from (and prepare for) extreme weather events.

Proactive adaptation, however, is the ultimate goal of the climate change research and practice communities. It is the counterpoint of reactive adaptation, in that adaptation can be proactive in either stimulus, wherein the driving force behind the action taken is the prediction or expectation of events not yet experienced; or form, including adaptive responses, such as the building of flood-protection infrastructure, that prevent the impact from taking a toll. As mentioned above, insurance is a more recent example of reactive adaptation to climate. Insurance companies are only beginning to integrate predictions based on an understanding of anthropogenic climate change into their insurance schemes, which would then be considered adaptation that is proactive in stimulus, but reactive in form.

Proactive adaptation is much more complicated than reactive adaptation because expected impacts of climate change are only now becoming clearer. Although our knowledge is constantly improving, infrastructure decisions depend on accurate information about the extent, timing, and distribution of impacts. Without this information, it is difficult to build a case for new flood-management strategies or approaches to city planning, for instance. As such, proactive adaptation is the frontier of climate change responses, and requires significant advances in the science of climate change. Particularly crucial will be better regional modeling of climate change impacts, so that we can estimate, for instance, how often and how severe flooding might be in a particular area, and a better understanding of the adaptation options at our disposal. Regional models are extremely challenging to formulate with a high degree of certainty, however, and significant advances must be made before these models are as robust and as applicable as global models.

Although humans have experience with adapting to climate, anthropogenic climate change is pushing communities outside their comfort zone. Not only are droughts and heat waves more extreme in many places than have been seen in the past, but new impacts are emerging, such as the increasingly rapid rise of sea levels and the deepening unpredictability of climate change. As such, the ability to adapt proactively to climate change is constrained by knowledge of the phenomenon, as well as by the political will to take precautionary action. Thus, adaptive actions are rarely taken in response to climate change alone, but in response to a number of converging impacts and vulnerabilities.

Adaptation strategies generally can be slotted into one or more of four categories. First, from the perspective of the natural environment, we can use knowledge of particularly sensitive and exposed ecosystems to impose stricter conservation and preservation guidelines. This might help to preserve ecosystems or species that are deemed to be especially valuable, although this also requires rather contentious decisions.

Second, we can address the underlying drivers of vulnerability. To some extent, official development assistance – international aid wealthy countries provide – and humanitarian programs have been doing this for decades, but we could (and are beginning to) weave adaptation to climate change into these plans to prioritize regions and communities that are especially vulnerable. Research conclusively shows that literacy, nutrition, education, and security of livelihoods are important components of resilience to climate change.

The third category of climate change adaptation responses pertains directly to projected impacts. As communities explore future climate change scenarios, they gain a better idea of the actions they need to take to protect themselves against potentially catastrophic events. City planning that promotes distributed energy generation, a secure food supply, and protected vital infrastructure falls into this category.

Finally, we simply need to understand more about the causes and consequences of climate change. We need greater certainty in our scenarios and a better understanding of how human systems will respond to predicted changes. As such, the science (and social science) of climate change must continue to improve to support adaptation.

To sum up, understanding the likely consequences of climate change will help us to manage exposure to the phenomenon and to build a case for more climate change mitigation. Giving decision-makers the technological tools and informational resources they need to develop adaptation plans will help to enhance communities' adaptive capacity. Finally, integrating adaptation into existing policies and programs will help to address the problem of finite resources and competing policy pressures. Together, these strategies will enhance the resilience of communities and could minimize climate change impacts over time.

10.5. Policy tools and progress

10.5.1. Policy tools for adaptation

Intentional, or proactive, adaptation to climate change is a much newer topic of negotiation than mitigation in the climate change community. Policy frameworks for adaptation are even rarer, for several reasons. First, as we have seen, the impacts of climate change are often most severe in resource-poor developing countries. In some cases, adaptation consists of relatively simple strategies with which communities have extensive experience, such as irrigation, building houses on stilts, and rotating crops to take advantage of precipitation patterns. In other cases, however, new technologies will be required, as well as detailed information about expected climate change impacts.

So, as with mitigation, many parties are advocating for an international framework in which developed countries financially support coordinated adaptation in developing countries.

This requires acknowledgment of the need for adaptation among developed countries, as well as willingness to fund initiatives that have mostly local benefits. This latter point is a common element of adaptation initiatives, which are often local in both origin and funding – keeping in mind the relatively blurry nature of the distinction between local and global, as described above. Nonetheless, as we shall see, efforts to design and implement adaptation strategies at the national level are gaining momentum.

Adaptation also differs from mitigation in that it is most often integrated with policies of other kinds. Development initiatives, disaster-risk management, conservation, and agricultural policies are just a few examples of the areas of pre-existing policy with which adaptation strategies must be integrated.

The design of adaptation policy is evolving, as our understanding of its influence on broader sustainable development priorities deepens. Nevertheless, a handful of core questions have structured adaptation efforts thus far:

- What are the existing methods for managing climate risks and adaptation? Are these viable in the future? Can they be built upon?
- What other interventions can be used to reduce climate change impacts and improve development outcomes?
- What are the costs, effects, and barriers of each option (based on agreed criteria)?
- How do the options compare through ranking?
- What suite of policies and measures constitutes a cohesive approach to development and adaptation?[15]

To take advantage of existing skills, technologies, and institutions, adaptation should build on existing methods for managing climate risks. As we have discussed, humans have adapted to climate for millennia and, in many cases, communities have effective systems in place for ameliorating risks. Climate change challenges this capacity, however, so new knowledge and technologies must be added to past experience.

Next, we must consider the range of available interventions. For example, if flooding is a problem for a low-lying community, it can consider a dike system, green berms, or wholesale withdrawal of the community from the vulnerable area. These options then must be evaluated in terms of barriers (such as the public perception of leaving the familiar space), costs, and effects, which, in turn, should be used to evaluate options and build a suite of policies that will accomplish the chosen intervention.

10.5.2. International and national adaptation

Momentum is now building behind a global effort to accelerate adaptation to climate change in a way that mirrors the mitigation targets countries have agreed to in the Kyoto Protocol of the United Nations Framework Convention on Climate Change (UNFCCC). For example, parties to the Kyoto Protocol have established an Adaptation Fund to finance concrete adaptation projects and programs in developing countries that are also parties to the Protocol. The Global Environment Facility, located at the World Bank headquarters in Washington, DC, provides its services to the Adaptation Fund as secretariat.

Indeed, lack of funding is a significant constraint on the world's most vulnerable countries' capacity to adapt to climate change. Although meaningful sums have been pledged to adaptation through the UNFCCC process, not much has yet been made available. Some progress was made during the negotiations at the Copenhagen Climate Change Conference in 2009, with the US Secretary of State Hillary Clinton promising to help raise US$100 billion per year by 2020 to help pay for both adaptation and mitigation in developing countries. Unfortunately, this proposed transfer of funds is dwarfed by the true costs of the impacts of climate change, so conversations continue about the most equitable way to address issues of responsibility and vulnerability. As well, the United Nations' "National adaptation programmes of action (NAPAs) provide a process for Least Developed Countries … to identify priority activities that respond to their urgent and immediate needs to adapt to climate change – those for which further delay would increase vulnerability and/or costs at a later stage."[16] As of 2012, the UNFCCC had received NAPAs for forty-seven countries,[17] and these plans are now being implemented.

10.5.3. Subnational adaptation

Climate change is affecting local communities in different ways, so action to address this challenge also must take place at the local level. Countries are recognizing the importance of pre-emptive action to address their particular vulnerability to climate change, and many have begun to address adaptation concerns, either within broader state climate action plans or through separate efforts matching their mitigation activities. As climate adaptation gains greater attention and resources, countries will have much to learn from places where adaptation is already occurring.

In California, for instance, political leaders have recognized that climate change is having a wide range of impacts on the state's natural resources, ecosystems, infrastructure, health systems and economy. In June 2005, Governor Arnold Schwarzenegger signed an executive order calling for biannual updates on the impacts of climate change on California, as well as adaptation plans to address these impacts. In 2008, the governor ordered state agencies to plan for sea-level rise and climate impacts. As climate change accelerates, it will strain systems further – bringing hotter and drier summers, increased risk of drought and wildfires, and expanded water resource needs. Through the California Energy Commission's Public Interest Energy Research program (PIER), research is under way to identify effective adaptation methods for agriculture, water resources and supply management, forest resources and wildfire management, and public health.

One community in Sweden is taking ecosystem-based approaches to climate change adaptation very seriously as part of a broader community transition towards sustainability. The neighborhood of Augustenborg, in the city of Malmö, was plagued by floods from overflowing stormwater systems and facing socio-economic decline. Rather than simply building a bigger set of pipes, Augustenborg is using what it calls a Sustainable Urban Drainage System to deal with increased runoff and waste management. A system of wetlands and drainage swales is currently in place, which is managing storm water while also sinking carbon and providing a beautiful landscape for residents to use. It is also helping to preserve habitat for bird and amphibian species, at significantly lower costs than traditional infrastructure.

Another example of adaptation is taking place in the Maldives, an island nation in the Indian Ocean and the planet's lowest-lying country. Around 330,000 people live on 192 islands, many of which are exposed to significant wave action and the threat of rising sea levels. In response to this risk and also to the spectacular damage that results from tsunamis, the Maldivian government has begun constructing "safe" islands: literally the product of dredging sand from the ocean floor and building a new island with a larger buffer zone between settlements and the ocean. Renewable energy, waste management, and coral reef preservation are also part of the "safe islands" approach. Criticisms of the authenticity of this approach have arisen, however: the Maldives is heavily dependent on tourism, nearly all of which arrives via airplane. Although any efforts to spur community adaptation and the uptake of local renewable energy are laudable, it is clear that fossil fuel consumption strictly within the borders of the country is not the only way that this idyllic island paradise is participating in growing greenhouse gas emissions.

Although adaptation measures are being implemented in communities and countries around the world, coordinated, proactive climate change adaptation supported by monitoring and sufficient funding remains rare. As the pace of climate change accelerates, however, this conversation will intensify, highlighting the importance of opportunities to share successful approaches.

10.5.4. Social movements and human behavior: The root of the adaptation conundrum

If we have a pretty good idea of some of the impacts of climate change, and we know that they are only getting worse, why are we not acting? The answer to this question gets at the roots of human behavior.

One reason adaptation around the globe is largely reactive and modest in scale is political. By declaring adaptation a priority, the perceived implication is that humans are having a discernable impact on climate and that mitigation has either failed or is insufficient to deal with the problem. Both of these claims are extremely politically contentious, and reveal the need to address such fundamental issues as the structure of the global energy system, patterns of inequality that leave millions of people ill-equipped to respond to climate change that, one could argue, they did not cause, and the ultimate implications of an economic system that is predicated on endless growth. In nations that are well-equipped with resources, information, and legitimate governance institutions, however, some would say that the real question is whether adaptation is proactive or reactive, not whether adaptation happens at all. Although effective mitigation – that is, reducing greenhouse gas emissions – is contingent on other large emitters simultaneously doing their part to reduce emissions, adaptation can be done relatively independently, unless problems emerge from issues such as rivers that flow across national boundaries.

Spurring adaptation to climate change is thus a tricky business. The most compelling argument for proactive climate change adaptation in developed countries might be that adaptation strategies can have significant co-benefits – in other words, communities can protect themselves against climate change while addressing other priorities, such as health, biodiversity, and water quality. For instance, constructing lush wetlands and placid retention ponds in cities might help neighborhoods deal with increasingly frequent flooding, while also providing new spaces for recreation, enhancing biodiversity, and reducing the costs of water treatment, as in Augustenborg. Planting urban forests creates shade that provides relief during heat waves, while also sequestering carbon and contributing to climate change mitigation. Every day, we are learning more about these win-win strategies, many of which are desirable even if we were to find out tomorrow that climate change is caused by an alien plot, rather than by increasing concentrations of greenhouse gases.

This idea of co-benefits, in fact, is the essence of sustainable development. Populations that are literate, healthy, and have access to education, food, and clean water are less vulnerable to climate change – and to any number of other risks. It is how development is done that becomes the crucial question. Are resilient, complete communities being created that rely on low-carbon sources of energy, are protected from flooding and drought, and able to engage meaningfully in decision-making about the future? These are the dimensions of true sustainability, and integral to climate change adaptation.

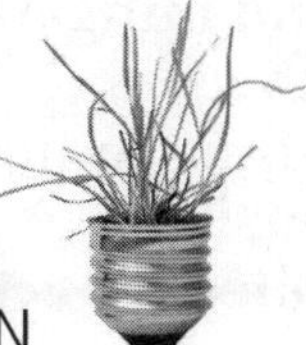

CHAPTER ELEVEN

Understanding Climate Change: Pathways Forward

MAIN POINTS:

- Pursuing sustainability, rather than just climate change, offers opportunities to achieve multiple priorities simultaneously.
- A development-path approach, or uncovering the underlying socio-economic, political, and technological drivers of both greenhouse gas emissions and vulnerability, could be a key ingredient in transformative change.
- Ethical dilemmas associated with climate change arise in part because the problem is global in nature, embodies intergenerational impacts, and deeply affects non-human nature, which is without a voice in global negotiations or local planning.
- Learning more, taking local action, and communicating widely about the complexities of climate change can trigger powerful shifts in the pathways communities around the world follow.

Earth's climate is steadily warming, and the human emission of greenhouse gases is largely responsible. Moreover, warming will continue even if we magically reduce emissions to zero tomorrow, so adaptation is both necessary and urgent. Despite compelling science and a suite of solutions that have been shown to be effective, climate change mitigation and adaptation have often been piecemeal, incremental, and ultimately insufficient to prevent further dangerous anthropogenic climate change.

So where does this leave us? Is the problem so overwhelming that action is futile? Exciting new trends in how we view the climate change problem suggest that sources of innovation exist that were previously ignored or underused. In this final chapter, we explore the power of framing climate change in the context of broader sustainability goals, the ethical challenges we face as we look at response options, and the potential for individual and collective action.

11.1. Integrating adaptation and mitigation: A sustainability approach

Inertia is built into the way we address climate change. Since the original formulation of the Intergovernmental Panel on Climate Change in 1988, the tasks of reducing greenhouse gas emissions and adapting to the impacts of climate change have been largely divorced from one another in the scientific and policy-making communities. This divorce was justified on the grounds that mitigation is largely global (the benefits of reducing a tonne of greenhouse gas anywhere on the globe benefits everyone), while adaptation is mostly local (the benefits of adaptation accrue mostly to the community in which the action is taken); or that mitigation is more technology driven or economic, while adaptation pertains to questions of development, poverty, and extreme events.

But are these distinctions as clear as they seem on the surface? A shift from coal-fired power plants to hydro power or solar would indeed reduce dramatically the greenhouse gas emissions that affect the entire planet. But it also would bring new technologies to local communities, the possibility of more decentralized (and thus more resilient in the face of extreme events) power production, improved air quality, and investment. Similarly, building a strong system of dikes will protect a community in Bangladesh from rising sea levels and powerful storms – clearly a direct benefit to a community whose livelihoods and health are threatened. But this adaptation also will prevent the creation of thousands of environmental refugees looking for new homes, and it might help to resolve local conflicts that can have global implications.

If we look at adaptation and mitigation through a systems lens, and think beyond their narrow first intention, we see a world in which responses to climate change can have unintended consequences and ripple effects. These ripple effects become even more significant if adaptation and mitigation are considered explicitly in tandem, or if the broader idea of sustainability, rather than climate change alone, is the focus.

Throughout this book, we have stressed the pressing need for creative, ambitious adaptation and mitigation. We have discussed ecosystem-based approaches – such as building wetlands to purify water and protect against floods – and renewable energy and the incredible opportunities (and challenges) these technologies offer. Many of these strategies solve only part of the problem, however, and are often pursued in relative isolation from one another. For example, a reduction in emissions might lead to increased vulnerability – picture a solar array, newly located on a low-lying and flood-prone area, replacing a coal-fired power plant farther inland, that is now vulnerable to sea-level rise and storms. If, however, communities plan for climate change by thinking first about *sustainability*, a curious thing can happen. The whole picture might become clearer: emissions come to matter as much as education; flood protection is explored alongside biodiversity conservation.

But what does "sustainability" actually mean? The classic definition arises out of the Brundtland Report, *Our Common Future*, published in 1987: "development that meets the needs of the present without compromising the ability of future generations to meet their own needs."[1] Over the past two and half decades, however, this definition has come under fire. If by development we mean economic growth, can this actually be sustainable in the long term? How do we define "needs"? If goals for development in Sweden differ from those in Ghana, can both be viewed as pursuing sustainable development?

So, a new way of viewing sustainable development – now often simply called sustainability, in part to shift the focus away from economic development that might not be desirable or feasible in the long term – is emerging. It is becoming clear that "sustainability" might mean something different to one person than it does to another – and that is acceptable.[2] Sustainability, after all, is about values: what a community wants to preserve, strengthen, and build – keeping in mind that the values of people in the future matter as well. What standard of living is considered desirable? Is this to be measured by looking at gross national product or by some other measure, such as gross national happiness? Should a community aim to preserve a mix of species that was present before it was built, or is a new pattern of habitats and species abundance acceptable? The point here is that creating and imposing a single definition of sustainability might reduce the likelihood that communities can envision a locally relevant, meaningful future that is motivating, feasible, and desirable. Despite the fruitful and exciting variety of sustainability visions, most attempt to balance several elements: ecological integrity, economic resilience (if not growth, in the traditional sense), and social equity.

In other words, communities around the world have multiple priorities. The urgency of responding to climate change is very real, but so are problems related to water quality, poverty, sanitation, literacy, community livability, biodiversity, and social equity. The challenge then becomes one of systems: where should planners or policy-makers draw the boundaries around the problem they want to solve? Are they thinking about synergies (how, for instance, sinking carbon by planting trees can also provide shade on a hot day and preserve crucial habitat) and trade-offs (is the construction of a massive concrete dike system to protect against floods damaging habitat and creating new greenhouse gas emissions in the process)? Can adaptation strategies be designed instead that sink carbon, provide recreation opportunities that enhance public health, conserve biodiversity, and enrich a neighborhood? The answer is absolutely yes.

The ecosystem-based approaches we discussed in Chapter 9 get us part of the way there, but still leave out pieces of the equation. Some scholars and practitioners suggest that, unless climate change policies are embedded in deeper changes in the way we do business as a society, it will be prohibitively expensive and disruptive to achieve our climate goals.[3] This suggests the desirability of integrating climate policy with broader sustainability goals relating to economics, social dimensions, technology, and the environment.[4] So, according to this view, sustainability must be woven throughout a broad range of policy priorities. For example, community development practices, fiscal mechanisms, health and education policies, and arts and culture could be mobilized to enhance community resilience. Decisions about strategic land use, urban form, energy, and transportation planning, and the benefits versus costs of retrofitting existing public, commercial, and residential assets are all part of the mix.[5] As such, the initial goal is to identify the levers or tools that might be used to trigger a shift in what we could call the underlying "development path."[6]

11.2. Development paths and transformative change

The challenge of addressing global climate change – including reducing greenhouse gas emissions to 80 percent below 1990 levels by 2050 – suggests that current models of urban development, energy provision, and consumption will need to be altered dramatically. In fact, eco-efficiency[7] – creating more goods and services but with progressively less waste and pollution – and incremental change[8] might serve only to perpetuate unsustainable trajectories. A fundamental shift in socio-technical systems and trajectories might be exactly what is required to address the challenge of climate change.[9] How this is to be cultivated within the policy constraints of communities poses interesting questions about what policy, technology, and social tools exist to generate the enabling conditions for this type of transformative change.[10]

The community is the level at which decisions about energy and transportation infrastructure, services, forest and biodiversity protection, agro-fuels cultivation, stormwater infrastructure, and natural hazard and flood risk systems will play out.[11] All of the associated policies, some of which are political priorities, will be influenced deeply by climate goals and strategies, while also having broader implications for community sustainability. Integration thus offers opportunities to embed climate mitigation and adaptation goals into existing mandates and budgets within municipal organizational structures.[12] What would this look like in practice? Plate 6 shows a typical Canadian street, this one in Burnaby, British Columbia: low density, pretty unattractive, and easily navigated only by car. What would this same street look like if climate change adaptation and mitigation were tackled simultaneously to create a more fundamentally sustainable community? Plate 7 shows one way this community could look, in a future where local renewable energy, public transit, active transportation, local food provision, and a host of sustainable strategies have been implemented.

A development path consists of a series of trajectories – economic, political, social, and technological – that are buried just below the surface of our everyday lives. They shape, for instance, what areas of land are considered desirable for cities, who can afford to build there, how homes are protected in vulnerable areas, how we move through the world, how far we have to travel to get to work or school, and where we spend our leisure time. Very few of these decisions are shaped by what we traditionally recognize as "climate change policy." Even so, the choices we make influence levels of greenhouse gas emissions and our vulnerability to the impacts of climate change.

A development-path framing of climate change responses suggests that highly non-linear opportunities might emerge to push drivers of emissions or vulnerability over a tipping point and institute a shift that cascades beyond the community in which the initial action took place.[13] What does a development-path approach actually mean in practice? It means taking a longer-term view in our planning process – a two- or three-year electoral cycle just is not sufficient if we are making decisions that will reverberate through decades. Tackling climate change through the underlying development path also requires us to use systems thinking, a theme we have reiterated throughout this book. What do decisions about energy supply mean for biodiversity? How can we design literacy programs and emergency management procedures to enhance resilience in the face of storms and other extreme weather?

Reducing greenhouse gas emissions and managing the impacts of climate change in a more holistic way also means that we need to use adaptive processes. We are working with a complex system, and we are getting new information every day. Suppose we decide to build a district energy system designed to heat homes with wood waste instead of with boilers powered by natural gas. We need information about the costs (financial, human) of the project. Will maintenance costs be higher than expected? Who will pay for building retrofits? How many people are required to run the system? We will also need to learn constantly about the effectiveness of the district energy system. Where will the wood waste come from? Will it be harvested sustainably? Will forests be replanted so that net greenhouse gas emissions are zero? All of this information will help decision-makers when they consider if district energy systems should be implemented in other neighborhoods, if a new source of fuel should be used instead of wood waste, and what the implications are for affordability, community vitality, and economic development in the neighborhood. Without this constant flow of information, and regular opportunities to tweak the initiative, we can get settled into deeply dysfunctional pathways that are not much better than the ones we were on before.

Finally, a development-path approach to climate change requires us to pay careful attention to *integrated decision-making*. Policies are commonly designed by one level of government, or even one unit within a level of government, that are completely inconsistent with policies made at another level. Take, for instance, a municipality that is struggling to reduce the emissions over which it has the most direct control: those from services such as waste collection, police and fire, libraries, and community recreation facilities. The municipality might carefully replace incandescent light bulbs with compact fluorescent bulbs, convert streetlights to LED lights, and even connect the community ice rink and pool so the heat sucked out of the ice rink to keep it cool is fed into the pool complex to keep it warm. At the same time, however, a higher level of government might design a transportation plan that aims to widen highways, increase cross-border shipping, and build a new terminal on the city's airport. The municipal policy is diligently tinkering away at less than 3 percent of total community emissions, while the regional plan increases transportation emissions by 300 percent.

These kinds of policy inconsistencies happen every day, in part because branches of government, community groups, and even individuals often work in isolation from one another. Procedures are also rarely in place to look at all policies through a sustainability lens that might reveal how a seemingly desirable policy has deeply undesirable consequences in some other area that matters.

What we are talking about here is a different sort of policy altogether: an approach to climate change that tackles the roots of emissions and vulnerability, rather than the symptoms of the disease once they have already emerged. This is what a sustainability lens, with attention to the underlying development path, has the potential to accomplish.

11.3. Ethics, equity, and responsibility

The ethical issues raised by the challenge of climate change are abundant and defy simplification. Climate change embodies three main qualities that make it a particularly prickly challenge. First, it is global, so local actions have global consequences. Some people are exposed to life-threatening impacts, while others are not – and often those who suffer the most are not those who caused the bulk of the problem in the first place. Second, our profligate emission of greenhouse gases has intergenerational consequence; we are making decisions about the climate that will affect people fifty, a hundred, or two hundred years from now. Third, many of the impacts of climate change are visited above all on non-human nature – the thousands of species that will move, suffer, or even become extinct as a result of rising sea levels, increased temperatures, and changing precipitation patterns. It turns out that our ethical toolkit is not quite adequate when it comes to valuing, or giving a voice to, nature, so we are left struggling to articulate the most appropriate relationship between humans and the natural world.[14]

Although climate change presents a myriad of ethical conundrums, let us highlight two issues in particular that can serve to represent ethical puzzles that are likely to continue to flare up in the near future: responsibility for paying for adaptation, and the challenge of geoengineering.

Negotiations under the United Nations Framework Convention on Climate Change often return to an important issue: if wealthy industrialized nations are responsible for the bulk of historical greenhouse gas emissions, but developing nations are suffering some of the most severe effects, who should pay to support adaptation in developing countries? The answer gets complicated when we recall that many of the drivers of vulnerability to climate change are rooted in levels of development: poverty, illiteracy, health, education, and other factors. So, funds for adaptation look more and more like funds for development, and most industrialized nations have already made commitments to support developing nations through foreign aid. Many scholars, decision-makers, and civil society groups argue, however, that additional funds should be raised for adaptation according to responsibility for the cause of climate change – that is, the emission of greenhouse gases – and allocated based on who is most vulnerable.[15]

Another key ethical dilemma relates to geoengineering – large-scale and intentional attempts to manipulate elements of Earth's climatic system. Fertilizing oceans with iron, for instance, might cause massive algal blooms that suck up carbon and sink it down into the deep. Injecting tiny reflective particles into the upper atmosphere might limit the amount of incoming radiation and control warming.

The ethical dilemmas associated with geoengineering inevitably will become more contentious as time passes and climate change impacts intensify, and they come in at least three flavors. First, the relatively low cost and technical simplicity of some geoengineering strategies mean that a company or a single government could make a decision that affects the climate of the entire planet. Of course, one could argue that this is already happening – that the emission of greenhouse gases is a massive experiment that is already affecting the planet. Second, the climatic system is full of feedbacks and multifaceted relationships. So geoengineering might lead to unintended consequences – such as ecosystem impacts, or reaching warming or cooling thresholds that we did not know existed – the implications of which might fall more heavily on some shoulders than others. Finally, there is the element of "moral hazard": even if geoengineering is proffered as a cheap, easy, transformative climate change response strategy, what motivation exists to tackle the roots of the problem – the consumption, patterns of development, and social inequalities that have led to current emissions and vulnerabilities?[16]

But critiques of geoengineering run deeper than simple fears about whether a company or country will take unilateral action or whether unintended consequences might be triggered. At the core of the debate is a more fundamental philosophical question: if our addiction to technology and our casual disregard for nature has caused the climate change problem, is geoengineering simply another manifestation of this dysfunction?[17] This question gets to the heart of different perspectives about how to address climate change, and is just one reason various stakeholders, governments, and individuals often appear to be talking past one another when they debate the wisdom of climate change adaptation and mitigation.

11.4. Individual choice and collective action: Moving forward

A deeper and more nuanced understanding of the science of climate change can have at least three effects. It can make us feel overwhelmed and paralyzed when we grasp the full magnitude of the challenge of climate change and the incredible effort required to reverse it. It can inspire us to take action individually, and to attempt to influence collective decisions on the issues. It can also have no effect at all. One thing, however, is clear: more and better data about climate change and growing scientific literacy are *insufficient* to motivate behavior change on a grand scale. These are absolutely crucial ingredients, but more is required. What is this "more," and who is responsible for providing it?

11.4.1. Evidence-based decision-making and the science/policy interface

The connections between our understanding of Earth systems and the design and implementation of policy traditionally have been pretty weak and inconsistent. Convincing evidence exists that humans are interfering with the planet's reflectivity through land-use changes, and with the composition of the atmosphere through the emission of greenhouse gases. Humans are throwing off a planetary balance that has been the foundation of our success as a species and upon which the safety and sustainability of communities around the world currently depend.

Scientists, entrepreneurs, and practitioners are also coming up with exciting ideas – many of which have been applied successfully in practice for decades – about how to respond: models for urban planning that reduce the need for transportation, constructed wetlands that cushion the blow of incoming storms, biofuels that chew up organic waste and heat our homes. But are these ideas making it successfully to decision-makers, so that they can be supported, implemented, shared, and scaled up? This is the question of the science/policy interface: what mechanisms are available for feeding science into decision-making? who is allowed to communicate with decision-makers? are individuals and organizations equipped with the tools they need to engage meaningfully with people in positions of power, or are they silenced by illiteracy, lack of understanding of the way decisions are made, or judgments about gender, ethnicity, and education?

Issues of public participation in decision-making are especially relevant to problems of global environmental change.

National-level governments are the main entity for making enforceable collective decisions,[18] but individuals also contribute to decision-making in some political systems through participatory processes, which represent a means by which government, science, business, and civil society actors can form powerful partnerships.[19] Cooperative decision-making processes allow for the incorporation of different ways of knowing. For example, aboriginal elders who pass down oral histories of the changing seasons could bring this knowledge to the table along with the atmospheric physicist's models.

Challenges arise, however, in ensuring that all participants are equally skilled in the necessary methods of communication, and so are equally represented. For this reason, more space is being given to new methods of deliberation, such as story-telling or visualization, which might be better placed to produce robust, compelling outcomes.[20] But we still have a participatory dilemma: should citizen groups participate in science-centered decision-making even if they are improperly equipped, or should they risk complete disenfranchisement?[21]

Global climate change provides a unique challenge to participatory processes. For much of the history of the climate change debate, the issue has been cast in highly technical, scientific terms, thus excluding vast portions of the lay citizenry. In addition, the global and diffuse nature of the impacts of climate change can lead to differing attributions of blame, while mitigation options depend on cohesive global commitment, and thus rest on varying socio-cultural development paths. As such, some scholars argue that we must recognize that some actions and decisions will prove more effective if the power to decide and act resides at the local level.[22]

11.5. Next steps

Now that you, the reader, have wrestled with the fundamentals of reflectivity and the intricacies of the carbon cycle, delved into the mysteries of mitigation, and explored the drivers of adaptation, what is next? The first step is to keep learning. This book has provided a little taste of the causes and consequences of climate change, but along the way you might have encountered a topic that grabbed your attention more than others. Dig deeper. Check trends in climate-related data over time: new data are emerging on atmospheric CO_2 concentration, global average temperature, and sea-level change, with dramatic implications for communities around the world. Check local trends, too: keeping an eye on local sea-level change, rainfall, snow pack, and extreme events, for instance, will help you to add detail to your own picture of local climate change. Following climate change in the media is a great way to grasp the various motivations behind responses to climate change, or delaying action. If you have the opportunity, contribute to this conversation yourself, by writing a letter to the editor, a blog post, or even simply commenting on an article.

The next step is to get active. It is very likely, no matter where you live on Earth, that someone, somewhere, in your community is tackling climate change. This might mean learning about decisions that your city council is making about renewable energy projects or plans for public transit. It might mean connecting with the Transition Towns movement, a network of hundreds of communities around the world that are quietly transforming emissions trajectories and vulnerability through grassroots initiatives.[23]

Finally, communicate with others. Conversations with friends, family, and co-workers might spark new ideas about how climate change can be addressed in your community, and trigger insights about what climate change means to different people. Understanding seemingly contradictory perspectives on the issue is key to moving the conversation forward and translating science into meaningful action.

Ultimately, climate change represents both a monumental challenge and a striking opportunity. It is absolutely possible to reduce greenhouse gas emissions and protect communities against the impacts of climate change, while making them more beautiful, healthy, and resilient. This is the essence of transformative change, and the seeds have already been planted. It is the task of every individual, community, and nation to cultivate them.

Notes

1. Climate Change in the Public Sphere

1 Many people mark the publication of Rachel Carson's landmark book *Silent Spring* (New York: Houghton Mifflin, 1962) as the critical juncture at which the public began mobilizing on issues of environmental pollution and the costs of Western development.

2 See, for instance, D.H. Meadows, D.L. Meadows, J. Randers, and W.W. Behrens III, *The Limits to Growth* (New York: New American Library, 1972), the seminal report commissioned by the Club of Rome that explores the implications of an exponentially increasing human population paired with limited resources.

3 J. Robinson, "Squaring the Circle? Some Thoughts on the Idea of Sustainable Development," *Ecological Economics* 48, no. 4 (2004): 369–84.

4 Recently, Professor John Robinson of the University of British Columbia has developed this perspective on the damage of a conversation about sustainability and climate change that focuses exclusively on limits and human activities. See, for example, J. Robinson, T. Berkhout, A. Cayuela, and A. Campbell, "Next Generation Sustainability at the University of British Columbia: The University as Societal Test-Bed for Sustainability," in *Regenerative Sustainable Development of Universities and Cities: The Role of Living Laboratories*, ed. A. König (Cheltenham, UK: Edward Elgar, 2013).

5 Ibid.

6 R.K. Pachauri and A Reisinger, eds., *Climate Change 2007: Synthesis Report. Contribution of Working Groups I, II and III to the Fourth Assessment Report of the Intergovernmental Panel on Climate Change* (Geneva: IPCC, 2007).

7 M.T. Boykoff, and J.M. Boykoff, "Balance as Bias: Global Warming and the US Prestige Press," *Global Environment Change* 14, no. 2 (2004): 126–36.

8 T. Satterfield, C. Mertz, and P. Slovic, "Discrimination, Vulnerability, and Justice in the Face of Risk," *Risk Analysis* 24, no. 1 (2004): 115–29.
9 D.M. Kahan, E. Peters, M. Wittlin, P. Slovic, L. Larrimore Ouellette, D. Braman, and O. Mandel, G., "The Polarizing Impact of Science Literacy and Numeracy on Perceived Climate Change Risks," *Nature Climate Change* 2 (2012): 732–5.
10 A. Kollmuss and M. Agyeman, "Mind the Gap: Why Do People Act Environmentally and What Are the Barriers to Pro-environmental Behaviour?" *Environmental Education Research* 3 (2002): 239–60.
11 P. Slovic, M.L. Finucane, E. Peters, and D.G. MacGregor, "The Affect Heuristic," *European Journal of Operational Research* 177, no. 3 (2002): 1333–52.
12 J. Eaton, "Eleven nations with large fossil-fuel subsidies," National Geographic Daily News, 18 June 2102; available online at http://news.nationalgeographic.com/news/energy/2012/06/pictures/120618-large-fossil-fuel-subsidies/, accessed 25 January 2014.
13 Ratification is the formal confirmation or adoption of (in this case) an international treaty. Although a "simple signature" by a state indicates that it agrees with the content of the treaty, ratification means that the state is willing to be legally bound by the treaty. Ratification requires domestic support for adherence to the treaty, as well as domestic legislation to enforce it (since the treaty cannot be enforced at the international level).
14 "The outcome at Copenhagen was disappointing. But if we work hard, there is still a way forward," *Observer*, 20 December 2009; available online at http://www.theguardian.com/commentisfree/2009/dec/20/leader-copenhagen-accord, accessed 9 October 2013.
15 United Nations Framework Convention on Climate Change, "Cancun Climate Change Conference – November 2010" (Bonn, Germany: UNFCCC, 2013); available online at http://unfccc.int/meetings/cancun_nov_2010/meeting/6266.php, accessed 9 October 2013.

2. Basic System Dynamics

1 Much of the systems terminology we use throughout this chapter is from Donella Meadows' excellent and accessible book, *Thinking in Systems: A Primer* (White River Junction, VT: Chelsea Green Publishing, 2008), which is highly recommended for further reading about systems.
2 See M.J. Molina and F.S. Rowland, "Stratospheric Sink for Chlorofluoromethanes: Chlorine Atom-catalysed Destruction of Ozone," *Nature* 249 (1974): 810–12; for a historical review, see Susan Solomon, "Stratospheric

Ozone Depletion: A Review of Concepts and History," *Reviews of Geophysics* 37, no. 3 (1999): 275–316.

3 S.A. Cameron, J.D. Lozier, J.P. Strange, J.B. Koch, N. Cordes, L.F. Solter, and T.L. Griswold, "Patterns of Widespread Decline in North American Bumble Bees," *Proceedings of the National Academy of Sciences* 108 (2011): 662–7.

4 J.D. Sterman and L.B. Sweeney, "Understanding Public Complacency about Climate Change: Adults' Mental Models of Climate Change Violate Conservation of Matter," *Climatic Change* 80, nos. 3–4 (2007): 213–38.

5 As some readers undoubtedly know, scientists (and others) use the terms "positive" and "negative" feedback for amplifying and stabilizing feedbacks, respectively. Because "positive feedback" and "negative feedback" have meanings in our common lexicon that differ from their scientific meanings, we chose to use the more descriptive terms, "amplifying" and "stabilizing"; see R.C.J. Somerville and S.J. Hassol, "Communicating the Science of Climate Change," *Physics Today* 64, no. 10 (2011): 48–53. If you are already comfortable with the "positive" and "negative" terminology and prefer to use them, just recognize that others, even fairly advanced science students, might misinterpret you.

6 The sweating feedback stabilizes your temperature until the heat is so extreme, or you are so dehydrated, that your body's capacity to regulate temperature through sweating is overwhelmed.

7 For a review of "relative age effects," see S. Cobley, J. Baker, N. Wattie, and J. McKenna, "Annual Age-grouping and Athlete Development: A Meta-analytical Review of Relative Age Effects in Sport," *Sports Medicine* 39, no. 3 (2009): 235–56. This feedback loop has been publicized more broadly in M. Gladwell, *Outliers: The Story of Success* (New York: Little, Brown, 2008).

8 J. Rockström, W. Steffen, K. Noone, Å. Persson, F.S. Chapin III, E.F. Lambin, T.M. Lenton, et al., "A Safe Operating Space for Humanity," *Nature* 461 (2009): 472–5.

9 Credit to Phil Austin, Earth, Ocean & Atmospheric Sciences, University of British Columbia, for this way of thinking about climate feedbacks.

10 Meadows, *Thinking in Systems*, 14.

11 Many versions of this fable are available, using rice or wheat, and are often used in mathematics lessons.

12 S. Solomon, K. Rosenlof, R. Portmann, J. Daniel, S. Davis, T. Sanford, and G.-K. Plattner, "Contributions of Stratospheric Water Vapor to Decadal Changes in the Rate of Global Warming," *Science* 327, no. 5970 (2010): 1219–23.

13 Pakistan, National Disaster Management Authority, *Annual Report 2010* (Islamabad: NDMA, 2011); available online at http://ndma.gov.pk/Documents/Annual%20Report/NDMA%20Annual%20Report%202010.pdf, accessed 18 May 2012.

14 R.D. Knabb, J.R. Rhome, and D.P. Brown, "Tropical Cyclone Report, Hurricane Katrina, 23–30 August 2005" (Miami: National Hurricane Center, 2005, updated 2011); available online at http://www.nhc.noaa.gov/pdf/TCR-AL122005_Katrina.pdf, accessed 18 May 2012.

15 D. Coumou and S. Rahmstorf, "A Decade of Weather Extremes," *Nature Climate Change* 2 (2012): 491–6.

16 The use of "Hiroshima bombs" as a metric comes from both John Cook, founder of skepticalscience.com, and James Hansen, former director of the National Aeronautics and Space Administration, Goddard Institute for Space Studies.

17 See S. Levitus, J.I. Antonov, T.P. Boyer, O.K. Baranova, H.E. Garcia, R.A. Locarnini, A.V. Mishonov, et al., "World Ocean Heat Content and Thermosteric Sea Level Change (0–2000 m), 1955–2010," *Geophysical Research Letters* 39, no. 10 (2012).

18 9 GtonC is equal to about 33 Gton CO_2 (the carbon plus the oxygen atoms in CO_2).

19 Data from United States, Department of Energy, Carbon Dioxide Information Analysis Center; available online at http://cdiac.ornl.gov/ftp/Global_Carbon_Project/Global_Carbon_Budget_2012_v1.5.xlsx, accessed 30 August 2013.

20 About 2.13 GtonC equals about 1 ppm CO_2 in the atmosphere, so 10.4 GtonC would add about 4.9 ppm CO_2.

21 Data from United States, National Oceanic and Atmospheric Administration; available online at ftp://aftp.cmdl.noaa.gov/products/trends/co2/co2_gr_mlo.txt, accessed 23 August 2013.

22 J.T. Fasullo and K.E. Trenberth, "A Less Cloudy Future: The Role of Subtropical Subsidence in Climate Sensitivity," *Science* 338, no. 6108 (2012): 792–4.

23 Of course, it is likely that some fossil fuel burning was involved in growing and transporting the food to you, which complicates matters.

24 W.A. Kurz, C.C. Dymond, G. Stinson, G.J. Rampley, E.T. Neilson, A.L. Carroll, T. Ebata, and L. Safranyik, "Mountain Pine Beetle and Forest Carbon Feedback to Climate Change," *Nature* 452 (2008): 987–90.

25 See, for example, D.R. Foster, "Land-Use History (1730–1990) and Vegetation Dynamics in Central New England, USA," *Journal of Ecology* 80, no. 4 (1992): 753–71.

26 C. Körner, "Plant CO_2 Responses: An Issue of Definition, Time and Resource Supply," *New Phytologist* 172, no. 3 (2006): 393–411.

27 S. Park, P. Croteau, K.A. Boering, D.M. Etheridge, D. Ferretti, P.J. Fraser, K-R. Kim, et al., "Trends and Seasonal Cycles in the Isotopic Composition of Nitrous Oxide since 1940," *Nature Geoscience* 5, no. 4 (2012): 261–5.

28 M. Sturm, T. Douglas, C. Racine, and G.E. Liston, "Changing Snow and Shrub Conditions Affect Albedo with Global Implications," *Journal of Geophysical Research* 110, no. G1 (2005).

29 For a review, see P.K. Quinn and T.S. Bates, "The Case against Climate Regulation via Oceanic Phytoplankton Sulphur Emissions," *Nature* 480 (2011): 51–6.

30 A. Robock and J. Mao, "The Volcanic Signal in Surface Temperature Observations," *Journal of Climate* 8, no. 5 (1995): 1086–103.

31 T. Gerlach, "Volcanic versus Anthropogenic Carbon Dioxide," *Eos, Transactions, American Geophysical Union* 92, no. 24 (2011): 201–2.

32 See, for example, K.M. Walter Anthony, P. Anthony, G. Grosse, and J. Chanton, "Geologic Methane Seeps along Boundaries of Arctic Permafrost Thaw and Melting Glaciers," *Nature Geoscience* 5 (2012): 419–26.

33 R.A. Myers and B. Worm, "Rapid Worldwide Depletion of Predatory Fish Communities," *Nature* 423 (2003): 280–3.

34 See, for example, W.M. Denevan, "The Pristine Myth: The Landscape of the Americas in 1492," *Annals of the Association of American Geographers* 82, no. 3 (1992): 369–85.

35 These numbers are best estimates for the period from March 2000 to May 2004 from K.E. Trenberth, J.T. Fasullo, and J. Kiehl, "Earth's Global Energy Budget," *Bulletin of the American Meteorological Society* 90, no. 3 (2009): 311–23.

36 R. Knutti and G.C. Hegerl, "The Equilibrium Sensitivity of the Earth's Temperature to Radiation Changes," *Nature Geoscience* 1 (2008): 735–43.

37 In 2007, Working Group I of the IPCC gave a "likely" range for the equilibrium climate sensitivity between 2 and 4.5°C, with a best estimate of about 3°C; see Intergovernmental Panel on Climate Change, "Summary for Policymakers," in *Climate Change 2007: The Physical Science Basis. Contribution of Working Group I to the Fourth Assessment Report of the Intergovernmental Panel on Climate Change*, ed. S. Solomon, D. Qin, M. Manning, Z. Chen, M. Marquis, K.B. Averyt, M.Tignor, and H.L. Miller (Cambridge: Cambridge University Press, 2007). In 2013, Working Group I of the IPCC declined to state a central value for the equilibrium climate sensitivity, but gave a "likely" range from 1.5 to 4.5°C; see http://www.ipcc.ch/report/ar5/wg1/docs/WGIAR5_SPM_brochure_en.pdf, accessed 8 February 2014.

38 A. Schmittner, N.M. Urban, J.D. Shakun, N.M. Mahowald, P.U. Clark, P.J. Bartlein, A.C. Mix, and A. Rosell-Melé, "Climate Sensitivity Estimated from Temperature Reconstructions of the Last Glacial Maximum," *Science* 334 (2011): 1385–8.

3. Climate Controls: Energy from the Sun

1 The ratio of the surface area of a sphere to the surface area of a circle (disc) is 4:1. Notice that 1365/4 = 341.
2 Many people hold an incorrect mental model of the reason for seasons. See the excellent 1987 video "A Private Universe," produced by the Harvard-Smithsonian Center for Astrophysics; available online at http://www.learner.org/resources/series28.html.
3 Some people erroneously imagine that Earth's axial tilt over the year is aligned like the string of a tetherball going around the central pole. But if this were true, one hemisphere would always have summer and the other always have winter, which is not the case.
4 N.A. Krivova, L.E.A. Vieira, and S.K. Solanki, "Reconstruction of Solar Spectral Irradiance since the Maunder Minimum," *Journal of Geophysical Research* 115, no. A12 (2010).
5 For a review of cosmogenic isotope records and solar activity, see J. Beer, "Long-term Indirect Indices of Solar Variability," *Space Science Reviews* 94, no. 1–2 (2000): 53–66.
6 C. Fröhlich, "Solar Irradiance Variability since 1978," *Space Science Reviews* 125, no. 1–4 (2006): 53–65.
7 Krivova, Vieira, and Solanki, "Reconstruction of Solar Spectral Irradiance."
8 Multiple research groups have used observational measurements to calculate Earth's average temperature over time using slightly different approaches, and their results agree. There are no credible challenges to data showing twentieth-century global temperature rise.
9 Other climate controls, such as volcanic activity, also might have played a role in this cooling; see T.J. Crowley, "Causes of Climate Change over the Past 1000 Years," *Science* 289, no. 5477 (2000): 270–7.
10 For recent data, see National Aeronautics and Space Administration, Marshall Space Flight Center, "Solar Physics: The Sunspot Cycle," available online at http://solarscience.msfc.nasa.gov/SunspotCycle.shtml.
11 See, for example, G. Bond, B. Kromer, J. Beer, R. Muscheler, M.N. Evans, W. Showers, S. Hoffmann, et al., "Persistent Solar Influence on North Atlantic Climate during the Holocene," *Science* 294, no. 5549 (2001): 2130–6.
12 D.W. Keith, E. Parson, and M.G. Morgan, "Opinion: Research on Global Sun Block Needed Now," *Nature* 463 (2010): 426–7.

13 T.M. Lenton and N.E. Vaughan, "The Radiative Forcing Potential of Different Climate Geoengineering Options," *Atmospheric Chemistry and Physics* 9 (2009): 5539–61.

14 Keith, Parson, and Morgan, "Opinion."

4. Climate Controls: Earth's Reflectivity

1 S.G. Warren and W.J. Wiscombe, "A Model for the Spectral Albedo of Snow, II: Snow Containing Atmospheric Aerosols," *Journal of the Atmospheric Sciences* 37, no. 12 (1980): 2734–45.

2 NASA tracks surface conditions on Greenland with satellites. See, for example, Maria-José Viñas, "Satellites See Unprecedented Greenland Ice Sheet Surface Melt" (Greenbelt, MD: National Aeronautics and Space Administration, Goddard Space Flight Center, 2012), available online at http://www.nasa.gov/topics/earth/features/greenland-melt.html accessed 22 December 2012; and http://polarmet35.mps.ohio-state.edu/albedo/0-3200m_Greenland_Ice_Sheet_Reflectivity_Byrd_Polar_Research_Center.png, accessed 22 December 2012.

3 E. Rignot, I. Velicogna, M.R. van den Broeke, A. Monaghan, and J.T.M. Lenaerts, "Acceleration of the Contribution of the Greenland and Antarctic Ice Sheets to Sea Level Rise," *Geophysical Research Letters* 38, no. 5 (2011).

4 G. McGranahan, D. Balk, and B. Anderson, "The Rising Tide: Assessing the Risks of Climate Change and Human Settlements in Low Elevation Coastal Zones," *Environment and Urbanization* 19, no. 1 (2007): 17–37.

5 J. Church and N. White, "Sea-Level Rise from the Late 19th to the Early 21st Century," *Surveys in Geophysics* 32, no. 4–5 (2011): 585–602.

6 For a synthesis of links between vegetation and climate, see M. Claussen, "Late Quaternary Vegetation-Climate Feedbacks," *Climate of the Past* 5 (2009): 203–16.

7 G.B. Bonan, "Forests and Climate Change: Forcings, Feedbacks, and the Climate Benefits of Forests," *Science* 320, no. 5882 (2008): 1444–9.

8 J. Hansen, M. Sato, P. Kharecha, D. Beerling, R. Berner, V. Masson-Delmotte, M. Pagani, et al., "Target Atmospheric CO_2: Where Should Humanity Aim?" *Open Atmospheric Science Journal* 2 (2008): 217–31.

9 S.K. Satheesh and K. Krishna Moorthy, "Radiative Effects of Natural Aerosols: A Review," *Atmospheric Environment* 39, no. 11 (2005): 2089–110.

10 For summaries of the effects of aerosols on clouds, see Chapters 2 & 7 in S. Solomon, D. Qin, M. Manning, Z. Chen, M. Marquis, K.B. Averyt, M. Tignor and H.L. Miller, eds., *Contribution of Working Group I to the Fourth Assessment Report of the Intergovernmental Panel on Climate Change* (United Kingdom and New York, NY: Cambridge University Press, 2007).

11 This is called the "first indirect effect" of aerosols, as opposed to the "direct" effects of aerosols themselves reflecting incoming solar radiation.
12 These are called the "second indirect effects" of aerosols on clouds.
13 This is called the "semi-direct effect" of aerosols on clouds.
14 A.E. Dessler, "A Determination of the Cloud Feedback from Climate Variations over the Past Decade," *Science* 330, no. 6010 (2010): 1523–7.
15 J.T. Fasullo and K.E. Trenberth, "A Less Cloudy Future: The Role of Subtropical Subsidence in Climate Sensitivity," *Science* 338, no. 6108 (2012): 792–4.
16 See S. Solomon, D. Qin, M. Manning, Z. Chen, M. Marquis, K.B. Averyt, M. Tignor and H.L. Miller, eds., *Contribution of Working Group I to the Fourth Assessment Report of the Intergovernmental Panel on Climate Change* (Cambridge: Cambridge University Press, 2007), chap. 2.
17 International Energy Agency, *Key World Energy Statistics, 2012* (Paris: IEA, 2012), available online at http://www.iea.org/publications/freepublications/publication/kwes.pdf, accessed 31 August 2013.
18 S.J. Smith, J. van Aardenne, Z. Klimont, R.J. Andres, A. Volke, and S. Delgado Arias, "Anthropogenic Sulfur Dioxide Emissions: 1850–2005," *Atmospheric Chemistry and Physics* 11 (2011): 1101–16; and Z. Klimont, S.J. Smith, and J. Cofala, "The Last Decade of Global Anthropogenic Sulfur Dioxide: 2000–2011 Emissions," Environmental Research Letters 8, no. 1 (2013).
19 J. Hansen and L. Nazarenko, "Soot Climate Forcing via Snow and Ice Albedos," *Proceedings of the National Academy of Sciences* 101, no. 2 (2004): 423–8.
20 V. Ramanathan, P.J. Crutzen, J. Lelieveld, A.P. Mitra, D. Althausen, J. Anderson, M.O. Andreae, et al., "Indian Ocean Experiment: An Integrated Analysis of the Climate Forcing and Effects of the Great Indo-Asian Haze," *Journal of Geophysical Research* 106 (2001): 28371–98.
21 Solomon, Qin, Manning, Chen, Marquis, Averyt, Tignor, and Miller, *Contribution of Working Group I to the Fourth Assessment Report of the Intergovernmental Panel on Climate Change.*
22 For a thorough summary and comparison of geoengineering techniques, see The Royal Society, *Geoengineering the Climate: Science, Governance and Uncertainty* (London: The Royal Society, 2009); available online at http://royalsociety.org/uploadedFiles/Royal_Society_Content/policy/publications/2009/8693.pdf, accessed 30 September 2013.
23 N.E. Vaughan and T.M. Lenton, "A Review of Climate Geoengineering Proposals," *Climatic Change* 109, no. 3–4 (2011): 745–90.
24 The exception, of those described here, is decreasing soot on snow, which does address a root problem.

5. Climate Controls: The Greenhouse Effect

1 International Energy Agency, *Key World Energy Statistics, 2012.*

2 See T. Boden, B. Andres, and G. Marland, "Global CO_2 Emissions from Fossil-Fuel Burning, Cement Manufacture, and Gas Flaring: 1751–2010" (Oak Ridge, TN: Oak Ridge National Laboratory, Carbon Dioxide Information Analysis Center, 2013), available online at http://cdiac.ornl.gov/ftp/ndp030/global.1751_2010.ems; and C. Le Quéré, R.J. Andrew, L. Bopp, J.G. Canadell, P. Ciais, S.C. Doney, P. Friedlingstein, et al., "The Global Carbon Budget 1959–2011," available online at http://cdiac.ornl.gov/ftp/Global_Carbon_Project/Global_Carbon_Budget_2012_v1.5.xlsx, accessed 30 August 2013.

3 Intergovernmental Panel on Climate Change, "Summary for Policymakers," in *Climate Change 2013: The Physical Science Basis. Contribution of Working Group I to the Fifth Assessment Report of the Intergovernmental Panel on Climate Change*, ed. T.F. Stocker, D. Qin, G.-K. Plattner, M.M.B. Tignor, S.K. Allen, J. Boschung, A. Nauels, et al. (Cambridge: Cambridge University Press, 2013), available online at-http://www.ipcc.ch/report/ar5/wg1/docs/WGIAR5_SPM_brochure_en.pdf, accessed 8 February 2014.

4 In contrast, the hot Sun emits shortwave radiation, with a peak at visible wavelengths. This incoming radiation is not involved in the greenhouse effect.

5 A.A. Lacis, G.A. Schmidt, D. Rind, and R.A. Ruedy, "Atmospheric CO_2: Principal Control Knob Governing Earth's Temperature," *Science* 330 (2010): 356–9.

6 G.A. Schmidt, R.A. Ruedy, R.L. Miller, and A.A. Lacis, "Attribution of the Present-day Total Greenhouse Effect," *Journal of Geophysical Research: Atmospheres* 115, no. D20 (2010).

7 There are, however, industrial compounds, such as chlorofluorocarbons and sulfur hexafluoride, that do absorb and emit in the atmospheric window. Their concentrations are currently low enough that most radiation at wavelengths in the atmospheric window does make it through into space.

8 North Greenland Ice Core Project Members, "High-resolution Record of Northern Hemisphere Climate Extending into the Last Interglacial Period," *Nature* 431 (2004): 147–51.

9 D. Lüthi, M. Le Floch, B. Bereiter, T. Blunier, J.-M. Barnola, U. Siegenthaler, D. Raynaud, et al., "High-resolution Carbon Dioxide Concentration Record 650,000–800,000 Years Before Present," *Nature* 453 (2008): 379–82.

10 N.J. Shackleton, "Carbon-13 in Uvigerina: Tropical Rainforest History and the Equatorial Pacific Carbonate Dissolution Cycles," in *The Fate of*

Fossil Fuel CO_2 in the Oceans, ed. N. Andersen and A. Malahof (New York: Plenum, 1977).

11 This event was the Paleocene-Eocene Thermal Maximum. For more information, start with M. Pagani, K. Caldeira, D. Archer, and J.C. Zachos, "An Ancient Carbon Mystery," *Science* 314 (2006): 1556–7.

12 Some evidence suggests that this carbon release occurred within a couple of decades. See J.D. Wright and M.F. Schaller, "Evidence for a Rapid Release of Carbon at the Paleocene-Eocene Thermal Maximum," *Proceedings of the National Academy of Sciences* 110, no. 40 (2013): 15908–13.

13 Y. Cui, L.R. Kump, A.J. Ridgwell, A.J. Charles, C.K. Junium, A.F. Diefendorf, K.H. Freeman, et al., "Slow Release of Fossil Carbon during the Palaeocene-Eocene Thermal Maximum," *Nature Geoscience* 4 (2011): 481–5.

14 J.C. Zachos, U. Röhl, S.A. Schellenberg, A. Sluijs, D.A. Hodell, D.C. Kelly, E. Thomas, et al., "Rapid Acidification of the Ocean during the Paleocene-Eocene Thermal Maximum," *Science* 308 (2005): 1611–5.

15 There have been many efforts to estimate Earth's energy imbalance. A place to start is Trenberth, Fasullo, and Kiehl, "Earth's Global Energy Budget."

16 J. Hansen, L. Nazarenko, R. Ruedy, M. Sato, J. Willis, A. Del Genio, D. Koch, et al., "Earth's Energy Imbalance: Confirmation and Implications," *Science* 308 (2005): 1431–5.

17 Dessler, "Determination of the Cloud Feedback."

18 For example, T.A. Boden, G. Marland, and R.J. Andres, "Global, Regional, and National Fossil-Fuel CO_2 Emissions" (Oak Ridge, TN: Oak Ridge National Laboratory, Carbon Dioxide Information Analysis Center, 2013).

19 R.A. Houghton, "Carbon Flux to the Atmosphere from Land-Use Changes: 1850-2005," In *TRENDS: A Compendium of Data on Global Change* (Oak Ridge, TN: Oak Ridge National Laboratory, Carbon Dioxide Information Analysis Center, 2008). See also Le Quéré, Andrew, Bopp, Canadell, Ciais, Doney, Friedlingstein, et al., "Global Carbon Budget 1959–2011."

20 J.E. Kutzbach, W.F. Ruddiman, S.J. Vavrus, and G. Philippon, "Climate Model Simulation of Anthropogenic Influence on Greenhouse-induced Climate Change (Early Agriculture to Modern): The Role of Ocean Feedbacks," *Climatic Change* 99, no. 3–4 (2010): 351–81.

21 W.F. Ruddiman, "The Anthropogenic Greenhouse Era Began Thousands of Years Ago," *Climatic Change* 61, no. 3 (2003): 261–93.

22 R.F. Keeling, S.C. Piper, A.F. Bollenbacher, and S.J. Walker, "Monthly Atmospheric $^{13}C/^{12}C$ Isotopic Ratios for 11 SIO Stations," In *TRENDS: A Compendium of Data on Global Change* (Oak Ridge, TN: Oak Ridge National Laboratory, Carbon Dioxide Information Analysis Center, 2010),

available online at http://cdiac.ornl.gov/trends/co2/iso-sio/iso-sio.html, accessed 1 October 2013.

23 J.B. Miller, S.J. Lehman, S.A. Montzka, C. Sweeney, B.R. Miller, A. Karion, C. Wolak, et al., "Linking Emissions of Fossil Fuel CO_2 and Other Anthropogenic Trace Gases Using Atmospheric $^{14}CO_2$," *Journal of Geophysical Research: Atmospheres* 117, no. D08302 (2012).

24 A.C. Manning and R.F. Keeling, "Global Oceanic and Land Biotic Carbon Sinks from the Scripps Atmospheric Oxygen Flask Sampling Network," *Tellus* B 58, no. 2 (2006): 95–116.

25 H.D. Matthews, N.P. Gillett, P.A. Stott, and K. Zickfeld, "The Proportionality of Global Warming to Cumulative Carbon Emissions," *Nature* 459 (2009): 829–33.

26 H.D. Matthews, S. Solomon, and R. Pierrehumbert, "Cumulative Carbon as a Policy Framework for Achieving Climate Stabilization," *Philosophical Transactions of the Royal Society A: Mathematical, Physical and Engineering Sciences* 370, no. 1974 (2012): 4365–79.

6. Climate Change Mitigation: Reducing Greenhouse Gas Emissions and Transforming the Energy System

1 Alberta, Environment and Sustainable Resource Development, "Greenhouse Gas Reduction Program" (Edmonton, 2012), available online at http://environment.alberta.ca/01838.html, accessed 27 November 2012.

2 Canada, Environment Canada, "Greenhouse Gas Emissions Data" (Ottawa, 2012), available online at https://www.ec.gc.ca/indicateurs-indicators/default.asp?lang=en&n=BFB1B398-1, accessed 27 November 2012.

3 United States, Department of State, *U.S. Climate Action Report 2010* (Washington, DC: Global Publishing Services, 2010); and idem, Environmental Protection Agency, *Inventory of U.S. Greenhouse Gas Emissions and Sinks: 1990–2011* (Washington, DC: EPA, April 2013).

4 N.M. Bianco, F.T. Litz, K.I. Meek, and R. Gasper, *Can the U.S. Get There from Here? Using Existing Federal Laws and State Action to Reduce Greenhouse Gas Emissions* (Washington, DC: World Resources Institute, 2013).

5 International Energy Agency, *Key World Energy Statistics, 2012.*

6 E. Shove, "Converging Conventions of Comfort, Cleanliness, and Convenience," *Journal of Consumer Policy* 26, no. 4 (2003): 385–418.

7 Kollmuss and Agyeman, "Mind the Gap."

8 See, for example, A. Irwin and B. Wynne, *Misunderstanding Science? The Public Reconstruction of Science* (Cambridge: Cambridge University

Press, 1996); S. Jasanoff and B. Wynne, "Science and Decisionmaking," in *Human Choice and Climate Change*, vol. 1, *The Societal Framework*, ed. S. Rayner and E. Malone (Columbus, OH: Batelle Press, 1998); and B. Wynne, "Misunderstood Misunderstanding: Social Identities and Public Uptake of Science," *Public Understanding of Science* 1, no. 3 (1992): 281–304.

9 F.G. Kaiser and S. Wolfing, "Environmental Attitude and Ecological Behaviour," *Journal of Environmental Psychology* 19, no. 1 (1999): 1–19.

10 I. Ajzen and M. Fishbein, *Understanding Attitudes and Predicting Social Behaviour* (Englewood Cliffs, NJ: Prentice-Hall, 1980)

11 See Kaiser and Wolfing, "Environmental Attitude and Ecological Behaviour"; D.G. Karp, "Values and Their Effect on Pro-environmental Behaviour," *Environment and Behaviour* 28, no. 1 (1996): 111–33; and Kollmuss and Agyeman, "Mind the Gap."

12 Canada, Environment Canada, "National Inventory Report 1990–2010: Greenhouse Gas Sources and Sinks in Canada – Executive Summary" (Ottawa, 2012), available online at http://www.ec.gc.ca/publications/default.asp?lang=En&xml=A91164E0-7CEB-4D61-841C-BEA8BAA223F9.

13 David Suzuki Foundation. "Transportation Solutions" (Vancouver, 2012); available online at http://www.davidsuzuki.org/issues/climate-change/science/climate-solutions/transportation-solutions/, accessed 20 November 2012.

14 Renewable Energy Policy Network for the 21st Century, *Renewables 2013: Global Status Report* (Paris: REN21, 2013).

15 Ibid., 17.

16 Ibid.

17 A. Holm, L. Blodgett, D. Jennejohn, and K. Gawell, *Geothermal Energy: International Market Update* (Washington, DC: Geothermal Energy Association, 2010).

18 See J. Lund, B. Sanner, L. Rybach, R. Curtis, and G. Hellstrom, "Geothermal (Ground-source) Heat Pumps: A World Overview," *Geo-Heat Centre Quarterly Bulletin* 25, no. 3 (2004): 1–10; and United States, Department of Energy, Federal Energy Management Program, "Geothermal Resources and Technologies" (Washington, DC, 2012); available online at http://www1.eere.energy.gov/femp/technologies/renewable_geothermal.html.

19 Worldwatch Institute, "Use and Capacity of Global Hydropower Increases" (Washington, DC, 2013); available online at http://www.worldwatch.org/node/9527, accessed 30 August 2013.

20 Other ways to enhance the land's capacity to sequester carbon are the use of soil amendments, and no-till agricultural practices.

7. Climate Models

1 P.N. Johnson-Laird, *Mental Models: Towards a Cognitive Science of Language, Inference, and Consciousness* (Cambridge, MA: Harvard University Press, 1983).
2 The different approaches and assumptions themselves are models – ways to represent the world realistically with less-than-perfectly-complete information.
3 Intergovernmental Panel on Climate Change, *Climate Change 2007, Synthesis Report.*
4 More than 3,600 drifters in the ARGO project measure temperature and salinity of the top 2,000 m of the oceans, as well as gather information about ocean currents. See http://www.argo.ucsd.edu/, accessed 2 October 2013.
5 Levitus et al., "World Ocean Heat Content."
6 See, for example, K.R. Gurney, R.M. Law, A.S. Denning, P.J. Rayner, D. Baker, P. Bousquet, L. Bruhwiler, et al., "Towards Robust Regional Estimates of CO_2 Sources and Sinks using Atmospheric Transport Models," *Nature* 415, no. 6872 (2002): 626–30.
7 See, for example, G.-K. Plattner, R. Knutti, F. Joos, T.F. Stocker, W. von Bloh, V. Brovkin, D. Cameron, et al., "Long-Term Climate Commitments Projected with Climate–Carbon Cycle Models," *Journal of Climate* 21, no. 12 (2008): 2721–51.
8 See, for example, D.J. Rowlands, D.J. Frame, D. Ackerley, T. Aina, B.B.B. Booth, C. Christensen, M. Collins, et al., "Broad Range of 2050 Warming from an Observationally Constrained Large Climate Model Ensemble," *Nature Geoscience* 5 (2012): 256–60.
9 For a thorough and accessible description and history of climate modeling, see P.N. Edwards, *A Vast Machine: Computer Models, Climate Data, and the Politics of Global Warming* (Cambridge, MA: MIT Press, 2010).
10 For descriptions of a variety of EMICs, see D.A. Randall, R.A. Wood, S. Bony, R. Colman, T. Fichefet, J. Fyfe, V. Kattsov, et al., "Climate Models and Their Evaluation," in *Climate Change 2007: The Physical Science Basis. Contribution of Working Group I to the Fourth Assessment Report of the Intergovernmental Panel on Climate Change*, ed. S. Solomon, D. Qin, M. Manning, Z. Chen, M. Marquis, K.B. Averyt, M.Tignor, and H.L. Miller (Cambridge: Cambridge University Press, 2007).
11 F. Giorgi, "Regional Climate Modeling: Status and Perspectives," *Journal de Physique* 139 (2006): 101–18.

12 For a discussion of regional climate modeling, see J.H. Christensen, B. Hewitson, A. Busuioc, A. Chen, X. Gao, I. Held, R. Jones, et al., "Regional Climate Projections," in *Climate Change 2007: The Physical Science Basis. Contribution of Working Group I to the Fourth Assessment Report of the Intergovernmental Panel on Climate Change*, ed. S. Solomon, D. Qin, M. Manning, Z. Chen, M. Marquis, K.B. Averyt, M.Tignor, and H.L. Miller (Cambridge: Cambridge University Press, 2007).

13 N. Stern, *Stern Review on the Economics of Climate Change* (London: HM Treasury, 2006); available online at http://web.archive.org/web/20061114045919/http://www.hm-treasury.gov.uk/independent_reviews/stern_review_economics_climate_change/stern_review_report.cfm, accessed 2 October 2013.

14 Randall et al., "Climate Models and Their Evaluation."

15 N. Oreskes, "The Role of Quantitative Models in Science," in *Models in Ecosystem Science*, ed. C.D. Canham, J.J. Cole, and W.K. Lauenroth (Princeton, NJ: Princeton University Press, 2003), 13.

8. Future Climate: Emissions, Climatic Shifts, and What to Do about Them

1 P. Raskin, T. Banuri, G. Gallopin, P. Gutman, A. Hammond, R. Kates, and R. Swart, *Great Transition: The Promise and Lure of Times Ahead* (Stockholm: Stockholm Environment Institute, 2002).

2 N. Nakicenovic, J. Alcamo, G. Davis, H.J.M. de Vries, J. Fenhann, S. Gaffin, K. Gregory, et al., *Special Report on Emissions Scenarios* (Cambridge: Cambridge University Press, 2000).

3 Intergovernmental Panel on Climate Change, *Climate Change 2001*; idem, *Climate Change 2007*; idem, *Climate Change 2013*.

4 I. Castles and D. Henderson, "The IPCC Emission Scenarios: An Economic-Statistical Critique," *Energy and Environment* 14, no. 2–3 (2003): 159–85.

5 R.S.J. Tol, B. O'Neill, and D. van Vuuren, "A Critical Assessment of the IPCC SRES Scenarios," ENSEMBLE-based Predictions of Climate Change and their Impacts, Project GOCE-CT-2003-505539 (Brussels: European Commission, 2005).

6 R.H. Moss, J.A. Edmonds, K.A. Hibbard, M.R. Manning, S.K. Rose, D.P. van Vuuren, T.R. Carter, et al., "The Next Generation of Scenarios for Climate Change Research and Assessment," *Nature* 463 (2010): 747–56.

7 Updates to the IPCC's newest Assessment Report are available online at http://www.ipcc.ch/. Information specific to the Fifth Assessment

Report scenarios process can be found online at http://sedac.ipcc-data.org/ddc/ar5_scenario_process/index.html.

8 Solomon, Qin, Manning, Chen, Marquis, Averyt, Tignor, and Miller, *Contribution of Working Group I to the Fourth Assessment Report*.

9 G.P. Peters, R.M. Andrew, T. Boden, J.G. Canadell, P. Ciais, C. Le Quéré, G. Marland, et al., "The Challenge to Keep Global Warming below 2°C," *Nature Climate Change* 3, no. 1 (2013): 4–6.

10 S. Dessai, W.N. Adger, M. Hulme, J. Turnpenny, J. Kohler, and R. Warren, "Defining and Experiencing Dangerous Climate Change," *Climatic Change* 64, no. 1–2 (2004): 11–25.

11 Ibid.

12 Peters, Andrew, Boden, Canadell, Ciais, Le Quéré, Marland, et al., "Challenge to Keep Global Warming below 2°C."

13 J.D. Neelin, M. Munnich, H. Su, J.E. Meyerson, and C.E. Holloway, "Tropical Drying Trends in Global Warming Models and Observations," *Proceedings of the National Academy of Sciences* 103, no. 16 (2006): 6110–15.

14 G.A. Meehl, T.F. Stocker, W.D. Collins, P. Friedlingstein, A.T. Gaye, J.M. Gregory, A. Kitoh, et al., "Global Climate Projections," in *Climate Change 2007: The Physical Science Basis. Contribution of Working Group I to the Fourth Assessment Report of the Intergovernmental Panel on Climate Change*, ed. S. Solomon, D. Qin, M. Manning, Z. Chen, M. Marquis, K.B. Averyt, M.Tignor, and H.L. Miller (Cambridge: Cambridge University Press, 2007).

15 National Drought Mitigation Center, "U.S. Drought Monitor Data Archive" (Lincoln, NE, 2013); available online at http://droughtmonitor.unl.edu/DataArchive.aspx, accessed 15 October 2013.

16 Meehl, Stocker, Collins, Friedlingstein, Gaye, Gregory, Kitoh, et al., "Global Climate Projections."

17 "Why Seas Are Rising Ahead of Predictions: Estimates of Future Rate of Sea-level Rise May Be Too Low," *Science Daily*, 1 November 2012; available online at http://www.sciencedaily.com/releases/2012/11/121101153549.htm, accessed 15 October 2013.

18 Ibid.

19 S. Rahmstorf, G. Foster, and A. Cazenave, "Comparing Climate Projections to Observations up to 2011," *Environmental Research Letters*, 7, no. 4 (2012).

20 J.H. Christensen and O.B. Christensen, "Severe Summertime Flooding in Europe," *Nature* 421 (2003): 805–6.

21 T.N. Palmer and J. Räisänen, "Quantifying the Risk of Extreme Seasonal Precipitation Events in a Changing Climate," *Nature* 415 (2002): 514–17.

22 R. Clark, S. Brown, and J. Murphy, "Modeling Northern Hemisphere Summer Heat Extreme Changes and Their Uncertainties using a Physics

Ensemble of Climate Sensitivity Experiments," *Journal of Climate* 19. no. 17 (2006): 4418–35.

23 A. Fouillet, G. Rey, F. Laurent, G. Pavillon, S. Bellec, C. Guihenneuc-Jouyaux, J. Clavel, et al., "Excess Mortality Related to August 2003 Heat Wave in France," *International Archives of Occupational and Environmental Health* 80, no. 1 (2006): 16–24.

24 Meehl, Stocker, Collins, Friedlingstein, Gaye, Gregory, Kitoh, et al., "Global Climate Projections."

25 Ibid.

26 M. McDiarmid, "Greenland glacier melting 5 times faster than in 1990s," *CBC News*, 29 November 2012; available online at http://www.cbc.ca/news/politics/greenland-glacier-melting-5-times-faster-than-in-1990s-1.1194070, accessed 15 October 2013.

27 Meehl, Stocker, Collins, Friedlingstein, Gaye, Gregory, Kitoh, et al., "Global Climate Projections."

28 See L. Bizikova, S. Burch, J. Robinson, A. Shaw, and S.R.J. Sheppard, "Adaptation, Climate Change, and Uncertainty," in *Climate Change and Policy: The Calculability of Climate Change and the Challenge of Uncertainty*, ed. G. Gramelsberger and J. Feichter (Heidelberg: Springer, 2011); B. Kasemir, J. Jäger, C.C. Jaeger, and M.T. Gardner, eds., *Public Participation in Sustainability Science: A Handbook* (Cambridge: Cambridge University Press, 2003); and M. Manning, M. Petit, D. Easterling, J. Murphy, A. Patwardhan, H.-H. Rogner, R. Swart, et al., eds. "Workshop Report" (Intergovernmental Panel on Climate Change Workshop on Describing Scientific Uncertainties in Climate Change to Support Analysis of Risk and of Options, Maynooth, Ireland, 11–13 May 2004).

29 S. Dessai, and M. Hulme, "Does Climate Adaptation Policy Need Probabilities?" *Climate Policy* 4, no. 2 (2004): 107–28.

30 B.C. O'Neill, and N.B. Melnikov, "Learning about Parameter and Structural Uncertainty in Carbon Cycle Models," *Climatic Change* 89, no. 1–2 (2008): 23–44.

31 Bizikova, Burch, Robinson, Shaw, and Sheppard, "Adaptation, Climate Change, and Uncertainty."

32 Jasanoff and Wynne, "Science and Decisionmaking."

33 Nakicenovic, Alcamo, Davis, de Vries, Fenhann, Gaffin, Gregory, et al., *Special Report on Emissions Scenarios*.

34 J. Robinson, "Future Subjunctive: Backcasting as Social Learning," *Futures* 35, no. 8 (2003): 839–56; and R. Swart, P. Raskin, J. Robinson, "The Problem of the Future: Sustainability Science and Scenario Analysis," *Global Environment Change* 14, no. 2 (2004): 137–46.

35 R. Robinson, S. Burch, M. O'Shea, S. Talwar, and M. Walsh, "Envisioning Sustainability Pathways: Recent Progress in the Use of Participatory Backcasting Approaches for Sustainability Research," *Technological Forecasting and Social Change* 78 (2011): 756–68.
36 See A. Lovins, "Energy Strategy: The Road not Taken?" *Foreign Affairs* 55, no. 1 (1976): 65–96; and J. Robinson, "Energy Backcasting: A Proposed Method of Policy Analysis," *Energy Policy* 10, no. 4 (1982): 337–44.
37 K.L. Anderson, "Reconciling the Electricity Industry with Sustainable Development: Backcasting – a Strategic Alternative," *Futures* 33, no. 7 (2007): 607–23; M. Höjer and L.-G. Mattson, "Determinism and Backcasting in Futures Studies," *Futures* 32, no. 7 (2000): 613–34; and K.H. Dreborg, "Essence of Backcasting," *Futures* 28, no. 9 (1996): 813–28.
38 A. Carlsson-Kanyamaa, K.H. Dreborg, H.C. Moll, and D. Padovan, "Participative Backcasting: A Tool for Involving Stakeholders in Local Sustainability Planning," *Futures* 40, no. 1 (2008): 34–46; and J. Quist and P. Vergragt, "Past and Future of Backcasting: The Shift to Stakeholder Participation and a Proposal for a Methodological Framework," *Futures* 38, no. 9 (2006): 1027–45.

9. Impacts of Climate Change on Natural Systems

1 It is important to think, as well, through a geographic lens: Where are the most altered ecosystems located? How do they factor into local economies? For more information about these and other effects, see the reports of Working Group II of the Intergovernmental Panel on Climate Change, as well as regional assessments that might be carried out by your national government or environmental agencies in your area.
2 S.H. Schneider, S. Semenov, A. Patwardhan, I. Burton, C.H.D. Magadza, M. Oppenheimer, A.B. Pittock, et al., "Assessing Key Vulnerabilities and the Risk from Climate Change," in *Climate Change 2007: Impacts, Adaptation and Vulnerability. Contribution of Working Group II to the Fourth Assessment Report of the Intergovernmental Panel on Climate Change*, ed. M.L. Parry, O.F. Canziani, J.P. Palutikof, P.J. van der Linden, and C.E. Hanson (Cambridge: Cambridge University Press, 2007).
3 L. Sweetlove, "Number of Species on Earth Tagged at 8.7 Million," *Nature*, 23 August 2011; available online at http://www.nature.com/news/2011/110823/full/news.2011.498.html.
4 R. Hickling, D.B. Roy, J.K. Hill, R. Fox, and C.D. Thomas, "The Distributions of a Wide Range of Taxonomic Groups Are Expanding Polewards," *Global Change Biology* 12, no. 3 (2006): 450–5.

5 J. Wormworth and K. Mallon, *Bird Species and Climate Change: The Global Status Report*, version 1.0 (Brisbane: Climate Risk Europe, 2006).

6 S.N. Stuart, J.S. Chanson, N.A. Cox, B.E. Young, A.S.L. Rodrigues, D.L. Fischmann, and R.W. Waller, "Status and Trends of Amphibian Declines and Extinctions Worldwide," *Science* 306, no. 5702 (2004): 1783–6.

7 A.R. Blaustein, S.C. Walls, B.A. Bancroft, J.J. Lawler, C.L. Searle, and S.S. Gervasi, "Direct and Indirect Effects of Climate Change on Amphibian Populations," *Diversity* 2, no. 2 (2010): 281–313.

8 J.A. Pounds, M.R. Bustamante, L.A. Coloma, J.A. Consuegra, M.P.L. Fodgen, P.N. Foster, E. La Marca, et al., "Widespread Amphibian Extinctions from Epidemic Disease Driven by Global Warming," *Nature* 439 (2006): 161–7.

9 R.R. Nemani, C.D. Keeling, H. Hashimoto, W.M. Jolly, S.C. Piper, C.J. Tucker, R.B. Myneni, et al., "Climate-driven Increases in Global Terrestrial Net Primary Production from 1982 to 1999," *Science* 300, no. 5625 (2003): 1560–3.

10 Kurz, Dymond, Stinson, Rampley, Neilson, Carroll, Ebata, et al., "Mountain Pine Beetle."

11 S. Piao, P. Ciais, Y. Huang, Z. Shen, S. Peng, J. Li, L. Zhou, et al., "The Impacts of Climate Change on Water Resources and Agriculture in China," *Nature* 467 (2010): 43–51.

12 J. Wang, R. Mendelsohn, A. Dinar, J. Huang, S. Rozelle, and L. Zhang, "The Impact of Climate Change on China's Agriculture," *Agricultural Economics* 40, no. 3 (2009): 323–37.

13 See Arctic Climate Impact Assessment, *Impacts of a Warming Arctic: Arctic Climate Impact Assessment* (Cambridge: Cambridge University Press, 2004); J. Overpeck, K. Hughen, D. Hardy, R. Bradley, R. Case, M. Douglas, B. Finney, et al., "Arctic Environmental Change of the Last Four Centuries," *Science* 278, no. 5341 (1997): 1251–6; and M.C. Serreze, J.E. Walsh, F.S. Chapin III, T. Osterkamp, M. Dyurgerov, V. Romanovsky, W.C. Oechel, et al., "Observational Evidence of Recent Change in the Northern High-latitude Environment," *Climatic Change* 46, no. 1–2 (2000): 159–207.

14 K. Tape, M. Sturm, and C. Racine, "The Evidence for Shrub Expansion in Northern Alaska and the Pan-Arctic," *Global Change Biology* 12 (2006): 686–702.

15 R. Pearson, S. Phillips, M. Loranty, P. Beck, T. Damoulas, S. Knight, and S. Goetz, "Shifts in Arctic Vegetation and Associated Feedbacks under Climate Change," *Nature Climate Change* 3 (2013): 673–7.

16 C. Rosenzweig, G. Casassa, D.J. Karoly, A. Imeson, C. Liu, A. Menzel, S. Rawlins, et al., "Assessment of Observed Changes and Responses in Natural and Managed Systems," in *Climate Change 2007: Impacts, Adaptation*

and Vulnerability. Contribution of Working Group II to the Fourth Assessment Report of the Intergovernmental Panel on Climate Change, ed. M.L. Parry, O.F. Canziani, J.P. Palutikof, P.J. van der Linden, and C.E. Hanson (Cambridge: Cambridge University Press, 2007).

17 Ibid.

18 See, for instance, P.W. Lawson, E.A. Logerwell, N.J. Mantua, R.C. Francis, and V.N. Agostini, "Environmental Factors Influencing Freshwater Survival and Smolt Production in Pacific Northwest Coho Salmon (Oncorhynchus kisutch)," *Can. J. Fish. Aquat. Sci.* 61 (2004): 360–73.

19 P. Verburg, R.E. Hecky and H. Kling, "Ecological Consequences of a Century of Warming in Lake Tanganyika," *Science* 301 (2003): 505–7.

20 R. Simon, M.W. Holderied, C.U. Koch, and O. von Helversen, "Floral Acoustics: Conspicuous Echoes of Dish-shaped Leaf Attract Bat Pollinators," *Science* 333, no. 6042 (2011): 631–3.

21 A cap-and-trade system involves placing a limit on the amount of greenhouse gas emissions that a company may release into the atmosphere. Companies that emit less than this amount may sell the difference between their own emissions and cap in the form of "permits" to companies that have exceeded the cap. This provides a financial incentive to reduce emissions to a level below the cap, and allows the market to determine the value of these emissions permits.

22 R. Costanza, R. D'Arge, R. De Groot, S. Farber, M. Grasso, B. Hannon, S. Naeem, et al., "The Value of the World's Ecosystem Services and Natural Capital," *Nature* 387 (1997): 253–60.

23 To learn more about the Red List, see the Web site at http://www.iucnredlist.org.

24 C. Prip, T. Gross, S. Johnston, and M. Vierros, *Biodiversity Planning: An Assessment of National Biodiversity Strategies and Action Plans* (Yokohama: United Nations University, Institute of Advanced Studies, 2010).

25 The case studies that follow, describing ecosystem-based approaches in the United Kingdom and Belarus, were originally published as part of a report to the European Commission. See S. Naumann, G. Anzaldua, H. Gerdes, A. Frelih-Larsen, M. Davis, P. Berry, S. Burch, and M. Sanders, "Assessment of the Potential of Ecosystem-based Approaches to Climate Change Adaptation and Mitigation in Europe, Final Report to the European Commission, DG Environment" (Brussels: European Commission, DG Environment, 2011).

26 I. Dickey and R. Tinch, "Wallasea Island Economic Benefits Study" (London: Economics for the Environment Consultancy Ltd., 2008).

27 Naumann, Anzaldua, Gerdes, Frelih-Larsen, Davis, Berry, Burch, and Sanders, "Assessment of the Potential of Ecosystem-based Approaches."

10. Climate Change Impacts on Human Systems

1 J. Fowlie and G. Hoekstra, "WorkSafe B.C. orders inspections of all B.C. sawmills after second devastating explosion," *Vancouver Sun*, 25 April 2012.
2 Although it is important to assess the relative vulnerabilities of countries around the world and to determine where action first needs to be taken, vulnerability assessments are based heavily on criteria that can be contentious; as such, they can differ widely from one another.
3 N. Brooks, W.N. Adger, and P.M. Kelly, "The Determinants of Vulnerability and Adaptive Capacity at the National Level and the Implications for Adaptation," *Global Environmental Change* 15, no. 2 (2005): 151–63.
4 G. Yohe and R. Tol, "Indicators for Social and Economic Coping Capacity: Moving toward a Working Definition of Adaptive Capacity," *Global Environmental Change* 12, no. 1 (2002): 25–40.
5 United Nations General Assembly, *Universal Declaration of Human Rights*, 217 A (III) (1948). The formal identification of this right was based on a process that began in 2003; see United Nations Committee on Economic Social and Cultural Rights, "General Comment No. 15, The Right to Water," E/C.12/2002/11 (2003).
6 United Nations, Department of Economic and Social Affairs, *Millennium Development Goals Report* (New York, 2008).
7 Z.W. Kundzewicz, L.J. Mata, N.W. Arnell, P. Döll, P. Kabat, B. Jiménez, K.A. Miller, et al., "Freshwater Resources and Their Management," in *Impacts, Adaptation and Vulnerability. Contribution of Working Group II to the Fourth Assessment Report of the Intergovernmental Panel on Climate Change*, ed. M.L. Parry, O.F. Canziani, J.P. Palutikof, P.J. van der Linden, and C.E. Hanson (Cambridge: Cambridge University Press, 2007).
8 K.E. Trenberth, P.D. Jones, P. Ambenje, R. Bojariu, D. Easterling, A. Klein Tank, D. Parker, et al., "Observations: Surface and Atmospheric Climate Change," in *Climate Change 2007: The Physical Science Basis. Contribution of Working Group I to the Fourth Assessment Report of the Intergovernmental Panel on Climate Change*, ed. S. Solomon, D. Qin, M. Manning, Z. Chen, M. Marquis, K.B. Averyt, M.Tignor, and H.L. Miller (Cambridge: Cambridge University Press, 2007).
9 P. Lemke, J. Ren, R.B. Alley, I. Allison, J. Carrasco, G. Flato, Y. Fujii, et al., "Observations: Changes in Snow, Ice and Frozen Ground," in *Climate Change 2007: The Physical Science Basis. Contribution of Working Group I to the Fourth Assessment Report of the Intergovernmental Panel on Climate Change*, ed. S. Solomon, D. Qin, M. Manning, Z. Chen, M. Marquis, K.B. Averyt, M.Tignor, and H.L. Miller (Cambridge: Cambridge University Press, 2007).

10 W.E. Easterling, P.K. Aggarwal, P. Batima, K.M. Brander, L. Erda, S.M. Howden, A. Kirilenko, et al., "Food, Fibre and Forest Products," in *Impacts, Adaptation and Vulnerability. Contribution of Working Group II to the Fourth Assessment Report of the Intergovernmental Panel on Climate Change*, ed. M.L. Parry, O.F. Canziani, J.P. Palutikof, P.J. van der Linden, and C.E. Hanson (Cambridge: Cambridge University Press, 2007).
11 Ibid.
12 Intergovernmental Panel on Climate Change, "Summary for Policymakers," in *Managing the Risks of Extreme Events and Disasters to Advance Climate Change Adaptation*, ed. C.B. Field, V. Barros, T.F. Stocker, D. Qin, D.J. Dokken, K.L. Ebi, M.D. Mastandrea, et al. (Cambridge: Cambridge University Press, 2012).
13 H.L. Berry, K. Bowen, and T. Kjellstrom, "Climate Change and Mental Health: A Causal Pathways Framework," *International Journal of Public Health* 55, no. 2 (2010): 123–32.
14 J. Larsen, "Setting the Record Straight: More than 52,000 Europeans Died from Heat in Summer 2003, Plan B Updates" (Washington, DC: Earth Policy Institute, 2006).
15 United Nations Development Programme, *Adaptation Policy Frameworks for Climate Change: Developing Strategies, Policies and Measures*, ed. B. Lim, E. Spanger-Siegfried, I. Burton, E. Malone, and S. Huq (Cambridge: Cambridge University Press, 2005).
16 See the UNFCCC Web site at https://unfccc.int/national_reports/napa/items/2719.php.
17 United Nations Framework Convention on Climate Change, "NAPAs Received by the Secretariat" (Bonn, Germany); available online at http://unfccc.int/cooperation_support/least_developed_countries_portal/submitted_napas/items/4585.php.

11. Understanding Climate Change: Pathways Forward

1 World Commission on Environment and Development, *Our Common Future* (New York: Oxford University Press, 1987).
2 Robinson, "Squaring the Circle?"
3 T. Morita, N. Nakicenovic, and J. Robinson, "Overview of Mitigation Scenarios for Global Climate Stabilization Based on New IPCC Emission Scenarios (SRES)," *Environmental Economics and Policy Studies* 3, no. 2 (2000): 65–88; J. Robinson and D. Herbert, "Integrating Climate Change and Sustainable Development," *International Journal of Global Environmental Issues* 1 (2001): 130–48; and R. Swart, J. Robinson, and S.

Cohen, "Climate Change and Sustainable Development: Expanding the Options," *Climate Policy* 3, Supplement 1 (2003): S19–S40.

4 L. Bizikova, S. Burch, S. Cohen, and J. Robinson, "A Participatory Integrated Assessment Approach to Local Climate Change Responses: Linking Sustainable Development with Climate Change Adaptation and Mitigation," in *Climate Change, Ethics and Human Security*, ed. K. O'Brien, B. Kristoffersen, and A. St. Clair (Cambridge: Cambridge University Press, 2010); and J. Robinson, M. Bradley, P. Busby, D. Connor, A. Murray, B. Sampson, and W. Soper, "Climate Change and Sustainable Development: Realizing the Opportunity," *Ambio* 35, no. 1 (2006): 2–8.

5 J. Robinson, T. Berkhout, S. Burch, E. Davis, N. Dusyk, and A. Shaw, "Infrastructure and Communities: The Path to Sustainable Communities" (Victoria, BC: Pacific Institute for Climate Solutions, 2008).

6 A. Shaw, S. Burch, F. Kristensen, J. Robinson, and A. Dale, "Accelerating the Sustainability Transition: Exploring Synergies between Adaptation and Mitigation in British Columbian Communities," *Global Environmental Change* (forthcoming).

7 J. Korhonen, and T. Seager, "Beyond Eco-efficiency: A Resilience Perspective," *Business Strategy and the Environment* 17, no. 7 (2008): 411–19.

8 T. Könnöllä and G. Unruh, G. "Really Changing the Course: The Limitations of Environmental Management Systems for Innovation," *Business Strategy and the Environment* 16, no. 8 (2007): 525–37.

9 J. Rotmans, R. Kemp, and M. van Asselt, "More Evolution than Revolution: Transition Management in Public Policy," *Foresight* 3, no. 1 (2001): 15–31.

10 S. Burch, A. Shaw, A. Dale, and J. Robinson, "Triggering Transformative Change: A Development Path Approach to Climate Change Response in Communities," *Climate Policy* (forthcoming).

11 Shaw, Burch, Kristensen, Robinson, and Dale, "Accelerating the Sustainability Transition."

12 L. Bizikova, J. Robinson, and S. Cohen, "Linking Climate Change and Sustainable Development at the Local Level," *Climate Policy* 7, no. 4 (2007): 271–7.

13 Burch, Shaw, Dale, and Robinson, "Triggering Transformative Change."

14 S.M. Gardiner and L. Hartzell-Nichols, "Ethics and Global Climate Change," *Nature Education Knowledge* 3, no. 10 (2012): 5.

15 M. Grasso, "An Ethical Approach to Climate Change," *Global Environmental Change* 20 (2010): 74–81.

16 D. Scott, "Geoengineering and Environmental Ethics," *Nature Education Knowledge* 3, no. 10 (2012): 10.

17 D. Jamieson, "Ethics and Intentional Climate Change," *Climatic Change* 33, no. 3 (1996): 323–36.

18 J.S. Dryzek, *Deliberative Democracy and Beyond: Liberals, Critics, and Contestations* (Oxford: Oxford University Press, 2000).
19 J. Burgess, J. Clark, and J. Chilvers, "Going 'Upstream': Issues arising with UK Experiments in Participatory Science and Technology Assessment," *Sociologia e Politiche Sociali* 8, no. 2 (2005): 107–36.
20 O. Renn, T. Webler, and P. Wiedemann, *Fairness and Competence in Citizen Participation: Evaluating Models for Environmental Discourse* (Dordrecht, Netherlands: Kluwer Academic Publishers, 1995).
21 A. Irwin, *Citizen Science* (New York: Routledge, 1995).
22 T. O'Riordan, *Globalism, Localism, and Identity: Fresh Perspectives on the Transition to Sustainability* (London: Earthscan, 2001).
23 To learn more about Transition Towns, see http://www.transitionnetwork.org.

Index